LEAD S

Safety leadership has been riding the waves of general leadership theories since the end of the last century. While this is not surprising – leadership is leadership, right? – it also means research has been running into the same issues as general leadership research. Because what is a good leader, or what makes a good leader? Scholars have looked at 'great men' and successful men. They have studied traits of important leaders, their attitudes, ethics, morality. As hard as these things are to study, they are equally hard to communicate, or to train. So what about leadership behaviours? What do safety leaders practically do? In this book you will find answers to these questions. It basically boils down to: safety leaders LEAD. What this means or entails, you will read in this book. The writers will lead you through LEAD with examples, exercises and theory. Leading has never been simpler.

— Frank Guldenmund, Delft, the Netherlands

LEAD Safety

A Practical Handbook for Frontline Supervisors and Safety Practitioners

Tristan William Casey
Mark Anthony Griffin

CRC Press
Taylor & Francis Group
Boca Raton London New York

CRC Press is an imprint of the
Taylor & Francis Group, an **informa** business

First edition published 2020
by CRC Press
6000 Broken Sound Parkway NW, Suite 300, Boca Raton, FL 33487-2742

and by CRC Press
2 Park Square, Milton Park, Abingdon, Oxon, OX14 4RN

Library of Congress Cataloging-in-Publication Data

Names: Casey, Tristan William, author. | Griffin, Mark Anthony, author.
Title: Lead safety : a practical handbook for frontline supervisors and safety practitioners / Dr. Tristan William Casey and Prof. Mark Anthony Griffin.
Description: First edition. | Boca Raton, FL : CRC Press, [2020] | Includes bibliographical references and index.
Identifiers: LCCN 2020009244 (print) | LCCN 2020009245 (ebook) | ISBN 9780367861131 (paperback) | ISBN 9780367861148 (hardback) | ISBN 9781003016953 (ebook)
Subjects: LCSH: Industrial safety--Management. | Leadership
Classification: LCC T55 .C367 2020 (print) | LCC T55 (ebook) | DDC 658.3/82--dc23
LC record available at https://lccn.loc.gov/2020009244
LC ebook record available at https://lccn.loc.gov/2020009245

ISBN: 9780367861148 (hbk)
ISBN: 9780367861131 (pbk)
ISBN: 9781003016953 (ebk)

Typeset in Palatino
by Deanta Global Publishing Services, Chennai, India

Contents

Acknowledgements

We would like to thank the Office of Industrial Relations for their ongoing support for the development and dissemination of the LEAD model and associated tools across industry, as well as the numerous businesses that have provided access to their organisations to further our research program.

Authors

Tristan William Casey is a Lecturer at Griffith University's Safety Science Innovation Lab and co-founder of boutique consultancy 'The Culture Effect'. Dr Casey is an experienced Organisational Psychologist with extensive experience in work health and safety and has a particular interest in leadership and culture. Dr Casey teaches the Safety Leadership program at Griffith University alongside Prof Sidney Dekker and Dr Drew Rae and has a high research and industry profile, having done numerous presentations and keynotes at academic and applied conferences. Dr Casey is currently studying his PhD, involving the LEAD model, with Profs Andrew Neal and Mark Anthony Griffin. Recently, Dr Casey's work was acknowledged publicly, taking out the NSCA 'Pinnacle' and 'Best Safety Leadership' awards with Teys Australia, and acknowledged as a semi-finalist for the Queensland Community Achievement Awards, both in 2019.

Mark Anthony Griffin is the Director of the Future of Work Institute at Curtin University, Western Australia. Mark's research examines the link between individual and organisational capability in areas such as safety, leadership, wellbeing, and productivity. He has conducted large-scale collaborative projects with a range of industries including transport, health, education, energy, mining, and finance. He has developed assessment tools for use in these industries across Australia, Europe, the UK, the US, and Asia using multilevel and longitudinal data. Mark is currently Associate Editor for the *Journal of Applied Psychology*, previous Associate Editor of *Journal of Management*, and founding Associate Editor of *Organizational Psychology Review*. He is a Fellow of the US Society for Industrial and Organizational Psychology, Past Chair of the Research Methods Division of the US Academy of Management, and a past recipient of an Australian Research Council Future Fellowship.

1

Why LEAD?

Chapter Summary

There are many different views on what safety leadership is and isn't. Workers tend to think safety leadership is primarily about compliance, which is secured through close scrutiny and enforcement of rules and standards. But safety leadership is so much more. The LEAD model is an evidence-based approach to safety leadership that lines up specific practices with specific situations (Figure 1.1). LEAD is also compatible with emerging ideas about how to do safety 'differently', as well as holding onto the best of what has come before. Overall, LEAD is built on a compelling argument that safety leadership is essentially good leadership, just applied to safety-specific settings. Understand the LEAD model, and you will understand how to lead not only safety, but all other aspects of work as well.

When you ask workers 'what is safety leadership', they tend to say the same things. They say: 'reminding us about protocols to follow', 'insisting on compliance', 'going over safety regulations', and 'saying that safety is very important'. These are all examples of what 100 workers actually said when asked this question.

Safety is usually seen as a process of exerting control. A process of pushing back, slowing down, and 'toeing the line'. Mostly, safety is achieved using rules, standard processes, and safe working procedures. Add leadership into the mix and it becomes a process of *enforcing compliance* with such rules, processes, and procedures. Going a little further, safety leadership is described as care and concern for human life and the reduction of suffering.

Safety leadership, when done well, is just as much about freeing up and empowering your workers, as it is about making work predictable and holding people to account for fair standards of performance. Safety leadership is about striving and encouraging success under routine conditions and also about creating an environment in which people feel safe to speak up and take on-board the lessons learned. Safety leadership is *purposeful*. Safety

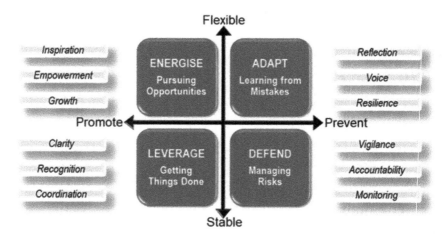

FIGURE 1.1
The full LEAD model of safety leadership.

leadership is *practical*. Safety leadership is *dynamic*. And most importantly, safety leadership is *specific to the situation*.

Here are some other ideas about the answer to the question: 'What is safety leadership?':

> Good safety leadership manages to integrate safety into overall organisational performance. So, it's not a standalone goal, but it's part of what they need to do as their job.

> I think probably the best way to describe safety leadership is that it's a focus on the safety culture of an organisation.
>
> It's about taking a proactive approach to engaging your people, and being interested in what's going on with those people.

> It's about interactions, and that's social interactions with people, and the way that in their social interaction that you convey the standard that you expect to see, or that you wish others to meet.

These quotes pave the way for exploration of a more complex, realistic, and, ultimately, more practical view of safety leadership.

LEAD takes safety leadership in exciting new directions, with practical tools to boot, that will give you the skills you need to make a difference to your team's safety. As a bonus, you will also see improvements in workers' engagement, morale, trust, and motivation to do the job well.

LEAD is designed to map specific safety leadership skills onto the work situation. This means that safety leadership is situationally specific – *what* you do and *when* you do it matters. It also means that effective safety leadership isn't a stubborn and fixed 'style' that we learn about and apply mercilessly across all work domains.

As explained in several scientific papers,[1] LEAD is a handy acronym that stands for four different safety leadership strategies: Leverage, Energise, Adapt, and Defend. Each of the strategies maps onto specific work situations. For example, Leverage is about acting to clarify goals and roles, reinforcing effective safety practices (e.g., giving recognition and reward), and coordinating work, all of which are best used when work is routine and low risk.

Energise is the next leadership strategy and refers to practices that concentrate on empowerment and ownership over decision-making among the team, inspiration, and fostering workers' growth and development. It is best used when change is being considered or implemented.

In contrast, Adapt is about creating flexibility and prevention, through practices such as encouraging workers to reflect on past performance, building resilience to emergency and high-tempo situations, and fostering a team environment where people feel safe to speak up and raise concerns. Consequently, Adapt is best used after an incident or following a major deployment or work shift.

Finally, Defend is the strategy that is best used during high-risk work. Practices include monitoring performance (i.e., taking an active interest in work), driving accountability and responsibility within the team, and creating a sense of vigilance and wariness to risks.

The four leadership strategies are important in all workplaces, so don't think of only one strategy as being the right one for you. Some workplaces might require more of one strategy than another. For example, working with hazardous chemicals will mean that Defend strategies are critical. However, we have found that a combination of strategies works best in almost every workplace. The reason is that every workplace will require people at different times to be motivated, careful, cooperative, adaptive, or vigilant … the list goes on. No single strategy fits all the time, and no single workplace works in only one way.

Have you ever experienced a static and fixed approach to safety in your workplace? If you haven't, I'm sure someone you know has. All it does is damage the reputation and credibility of safety. Policing safety in an inflexible, intolerant, and, dare I say it, ignorant way erodes trust, conveys a sense of disrespect, and drives a wedge between the leader and the worker. There has to be a better way.

Consider your typical working day. Within that day, there might be a whole host of different tasks and work situations that you deal with. Yet, safety is stubbornly fixed. 'Enforce that rule'; 'apply this procedure'; 'watch those troublesome workers and make sure they are following the right way of doing things'.

Most of what we know and do regarding safety, and as a direct result, safety leadership, is founded in thinking that is roughly a century old.[2] It's called 'Taylorism' or 'Scientific Management'. Taylorism is the product of Fredrick Taylor, a mechanical engineer who wanted to improve industrial efficiency.

Around the time of Taylor, we were just coming off the tail end of the Industrial Revolution, a period of great change and innovation. Taylor strived to understand how work could be done faster, better, and safer. In trying to reach this noble goal, Taylor made some discoveries and developed some principles about work:

- There is 'one best way' to do work,
- We can discover this best way by breaking work down into components, studying it, optimising it, and then reassembling it, and
- The way to achieve maximum output (and safety) is to prescribe the one best way into a procedure and enforce compliance with that procedure.

Many of you will see the hallmarks of modern safety management within these ideas promoted by Taylor. Unfortunately, Taylorism also drove a wedge between the management and the workers, which persists to this day. Taylor advocated for hierarchical organisational structures and the division of labour into those who do the work and those who design/manage/supervise the work.

Today, we see signs of 'Taylorism on steroids'. Reports from big consulting firms like Deloitte state that over $250 billion in the Australian economy is spent on compliance with regulations, laws, and internal bureaucracy.[3] Leaders spend eight hours or more each week doing compliance paperwork. Safety researchers are screaming that such activities, at best, add little value to operational safety, and at worst, actually make things less safe.

Subsequently, we find that leaders in modern organisations tend to sit separately from workers who perform the 'hands-on' parts of the job. Leaders specify the one best way to do things, communicate that way to workers, and enforce compliance (or reinforce effective performance). Such a leadership style is best described as 'transactional', and we see this applied to safety wholeheartedly. Does your organisation routinely use punishment to 'fix' problems with safety? What about the use of terms such as 'violation' and 'non-compliance'?

Transactional leadership is rife in the safety world. Many people think that they can only show their commitment to safety by writing a prescription, monitoring compliance, and sharply redirecting non-conformance. Granted, some of these actions will be needed from time to time, but this is just the tip of the iceberg when it comes to safety leadership.

At this point, I encourage you to think about the answers to the following questions:

- How do you define safety?
- How do you achieve safety?
- How do you know you have achieved safety?

Jot down your answers; we'll come back to them shortly.

Others take a small step beyond the transactional and believe that safety leadership is about genuine care and concern. Psychologically, showing care and concern to workers is thought to build commitment and motivation by activating the 'norm of reciprocity'. This norm or unwritten social rule states that when we receive something nice, we feel obligated to return the favour. This characteristic has been hardwired into us to create socially harmonious groups. It serves a useful purpose, and for safety, when care is coupled with transactional safety leadership (rewarding and punishing safe/unsafe acts), it creates the conditions for improved compliance.

This idea of 'care' as a tangible expression of safety commitment nudges things forward a little, but still doesn't get us to the core of what safety leadership is about. Safety leadership as care and concern slightly expands our domain of practice to include mental health and wellbeing, which is helpful, but still lacks the punch we need to drive truly exceptional safety performance.

In short, safety leadership is very much determined by our ideas about what safety is (and isn't) and the techniques and tools we have used to achieve safety. For decades, the dominant approach to safety has been rooted firmly in the idea of safety being about 'preventing the bad stuff' and achieving this by developing lumbering bureaucracies that are a complex web of procedures and rules. A safety leader's job then becomes that of an enforcer or police officer.

In some organisations, workers are expected to comply with literally hundreds of different safety procedures. Organisations, in their quest for moral supremacy, and cynically, perhaps to satisfy their clients' demands, espouse slogans like 'Zero Harm'. Yet is all this just 'wind in the willows'? How does a leader lead 'Zero Harm' in practice? How can a leader him/herself possibly enforce compliance with procedures and rules that workers, let alone the leader, couldn't possibly memorise or apply?

But this is changing. Take this short example of 'safety leadership differently' as a case in point:

> They had bookshelves full of procedures, operating procedures, safety procedures, so there's this poor culture about compliance. Often, they were not even workable, or they applied to a previous version of the equipment that they had. Top management said, 'Look, we're going to turn this around. We're going to actually listen to all our team members, invite them to have their say in how they actually do the jobs on a day-to-day basis and document that and not just document that blindly into procedures. Can it be a job aid or does it have to be a step-by-step multiple pages document?' They did this with all their workers and had discussions about this how we will go forward in the way that we do this job. Everybody could buy into that. They were involved in it. I think it was very modern thinking on the part of that organisation.

Our ideas about safety are evolving. Some would say a revolution is quietly brewing and bubbling away. A few organisations have already embraced it; they call it things like 'Safety Differently', 'The New View', and 'Safety-II'. For our purpose, let's just call it 'SD' for short.

SD turns safety on its head, and as a result, our ideas about safety leadership as well. SD adds to the story by defining safety as the 'presence of the good stuff' or fostering positive capacities to succeed under varying conditions.

SD means we ask tough questions about why we are doing certain things, particularly if they don't add value (think of the Take-5 tick-and-flick checklist). SD wants leaders to go beyond simple reward and punishment models and lift their eyes up from the forest of procedures so they, and their teams, can see the beauty of the sky above, and all the possibilities for growth, development, and improvement that it brings.

Whereas our traditional ways of managing safety are founded on ideas like 'people are a problem to control', 'variability or non-conformance is a threat to be corrected', and 'safety is a bureaucratic accountability', SD takes safety into a new era with ideas like 'people are a solution to harness', 'variability is an opportunity to learn', and 'safety is an ethical responsibility'.

What does SD mean in practice? And importantly, what does it mean for safety leadership?

Well, for starters, SD means that leaders need to build trust with their teams. We need to create an environment that is ripe for learning. An environment where people feel safe to take interpersonal risks, like sharing a mistake, non-conformance, or 'near-miss', without fear of retribution, punishment, or ridicule.

SD means we need to understand the way work is actually done. Rather than coming up with fanciful ways of doing a job that rarely matches the actual practice (work as imagined), leaders need to get among the weeds, get their hands dirty, and jump down into the trenches (I've run out of clichés).

Safety leadership, under the SD philosophy, is about removing barriers, hierarchy, and constraints. It's about truly understanding what makes work, and workers, tick and doing more of the things that make it successful. It's about paying equal attention to things that go well and things that go wrong. It's about fostering expertise within a team, growing capabilities, and fostering a sense of purpose and meaning. Safety leadership goes beyond safety and connects us back to 'good leadership'.

Another example of safety leadership that exemplifies SD is given below:

> People on the ground were armed with the right information – they were owning a lot more of their safety stuff and that was led purely by a leadership group who at one point in time wouldn't even get out of the office, and all of a sudden they're doing the field once a month to three days at a time and they were driving this new agenda and then all of a sudden you see a completely different demeanour. On the project, people weren't worrying about the 24/7 roster, they were there delivering their jobs and they made a significant amount of money doing it and they got out on time, on budget, and both companies were happy.

From an SD perspective, safety leadership is dynamic, which means our mode of operation changes in response to the work situation. It's no longer appropriate to have blanket rules and processes that apply blindly in all situations. Sure, some order and stability are helpful, but the trick is knowing when and how. LEAD offers some ideas in this space.

We need to measure safety by the presence of positives, as well as the absence of negatives, which means getting to know your team extremely well, warts and all, and working with them to develop their capacities to succeed. It means an effective safety leader knows when to apply the brakes, and how hard, and also knows when to put their foot on the accelerator, and even let their passengers have a go at taking the driving wheel.

Do you think you can build trust with your team if you throw safety procedures down their throats regardless of the situation? If you focus on the absence of negatives and compliance with unworkable processes? If you weigh safety heavily when the pressure is off, but when the proverbial hits the fan (which is precisely when we need safety the most), you focus instead on production?

I'll ask you those same questions again, differently (no pun intended):

- How could you define safety using new words and ideas?
- How else could you achieve safety?

- What else could you do to find out if you have achieved safety?
- What does all this mean for how you might do safety leadership differently?

For safety leadership to remain relevant and useful, our ideas about it must also evolve and keep up with this progress of how safety is thought about, practised, and improved.

An effective safety leader is nimble. He or she knows when the situation requires accountability; when it requires vigilance; and when it requires monitoring. Even when a situation strikes that needs this type or mode of safety leadership, he or she has the tools to do the job respectfully and understands the need to be fair, just, and understanding.

Now, I know what you might be thinking. Safety leadership is not just about being 'soft', 'going easy', or 'doing the fluffy stuff'. Quite the contrary. LEAD demands that leaders will need to make their expectations known; that they will set challenging goals and drive their teams towards them; and that they will engage with workers directly and influence participation in the organisation's initiatives. Safety leadership is not for the faint-hearted. You must thirst for improvement, success, and operating discipline. A LEADer uses the best of what has come before (Safety-I) and integrates it with the latest thinking in safety science (Safety-II): a blend of compliance and standardisation with the benefits of empowerment and engagement.

Not only is LEAD practical, but it is built on a foundation of science and evidence. There is a lot of pseudoscience out there when it comes to leadership, safety, and safety leadership. Rest assured, the LEAD model was developed through an academic collaboration between three universities and has also been adopted by Workplace Health and Safety Queensland (a safety regulator in Australia). We've carried out important and ground-breaking research to verify that LEAD is a valid and reliable safety leadership model (if you want to know more about this work, see Appendices 2, 3, and 4). You're in good hands.

LEAD will also carry benefits outside safety and beyond. When a leader engages with workers meaningfully, helps them to learn, clarifies goals, and reinforces high performance, *all* domains of work activity receive a shot in the arm. This happens because LEAD defines safety leadership as good leadership practices applied in the domain of safety. In short, there really is no such thing as 'safety leadership'. There's no magical combination of behaviours that safety leaders do that sets them apart from everyday leaders. If there was, we'd have an infinite number of 'leaderships': production leadership, innovation leadership, quality leadership … Plus, we like to keep things simple.

Phew!

Are you convinced?

Please, read on and learn more about LEAD. I can promise you that if none of what we said above resonates, you'll find at least *one* of your own reasons why you could use LEAD in your workplace to improve safety leadership.

Reflection Questions

- How has your definition of safety leadership changed (or not) as a result of this chapter?
- What signs of the 'old view' or Taylorism with respect to safety can you see in your organisation?
- What are some of the 'pointless' safety activities undertaken in your organisation, and what could you do to either remove them or increase their effectiveness?

Practice Points

- Safety leadership is good general leadership applied to safety-relevant settings.
- Consider the specific work situation and how your leadership skills can maximise your team's performance in that situation.
- LEAD has four key bundles of practices: Leverage, Energise, Adapt, and Defend, which require markedly different skills to execute effectively.
- The key to good safety lies in embracing flexibility and adaptability as much as it does in ensuring compliance and minimising variability; the trick is in understanding when to be flexible and when to emphasise consistency.

References

1. Casey, T., Griffin, M. A., Flatau Harrison, H., & Neal, A. (2017). Safety climate and culture: integrating psychological and systems perspectives. *Journal of Occupational Health Psychology, 22*(3), 341.
2. Dekker, S. (2019). *Foundations of Safety Science: A Century of Understanding Accidents and Disasters*. Abingdon, UK: Routledge.
3. Deloitte (2014). Rules eat up $250 billion a year in profit and productivity. Retrieved from https://www2.deloitte.com/au/en/pages/media-releases/articles/rules-eat-up-250-billion-a-year-271014.html.

2

LEAD in a Nutshell

Chapter Summary

There is much more to a leader than charisma. In other words, leaders are made, not born. We each have our own unique approach to leadership, and LEAD provides a scaffold around which we can show our unique approach but also ensure some consistency in leadership practice. This LEAD scaffold is underpinned by some fundamental trade-offs that must be reconciled or 'solved'. The first is the prevention–promotion trade-off, which contrasts growth, achievement, and exploration against compliance, risk-aversion, and exploitation. The second is the flexibility–stability trade-off. This trade-off contrasts openness to change and pursuit of creativity against a preference for stability and order, as achieved through deferring to established processes and procedures. Each situation requires a different resolution of these trade-offs, which is achieved through specific leadership practices: each element of the LEAD model.

Let's cut straight to the chase. LEAD is based on a few core ideas. Safety leadership is:

- Dynamic, or in other words, fluid and constantly changing,
- Matched to the requirements of the situation,
- About resolving or reconciling important 'trade-offs', and
- Consists of four bundles of practices (Leverage, Energise, Adapt, and Defend).

Let's explore each of these ideas in turn.

A lot of leadership theory is based on the idea that it is detached from the work situation. Leadership tends to be described in ways that make it seem global and fixed: give rewards, apply punishments, present a vision, develop a genuine relationship, be charismatic and a 'hero' to inspire your people, a role model of what you expect to see ('walk the talk'), and inspire your team with positive words … need I say more?

Other ideas about leadership paint a picture that seems to depend heavily on the person. Some leadership theories talk about the 'charismatic leader'[1] – someone who displays an aura or air of confidence, someone who can rally the troops through their mere presence. The problem with this idea of leadership is that it is firmly rooted in personality. You either have it or you don't. And we know that some of the most uncharismatic people make the best leaders, in terms of business success.

A charismatic leader can also be detrimental to a team and an organisation. Charismatic leaders can discourage independent and creative thinking, instead preferring to take this task on themselves and creating a group of blind followers rather than proactive self-leaders. Charismatic leaders create a whole host of dysfunctional dependencies. For instance, what happens when the time comes for the charismatic leader to move on? Has he or she developed an adequate succession plan and fostered growth and development among their followers?

Another issue is that many existing theories of leadership fall short in terms of their translation into actions: Develop a vision of what exactly? Role model what types of behaviours, and when? What sort of safety behaviours should be reinforced, and which should be discouraged?

The problem with all these ideas is that they aren't very practical. Neither are they accurate of what leaders actually *do*. Real leadership, and indeed, real safety leadership, is a complex toing and froing as work is done, mistakes are made, threats are repelled, and opportunities seized.

LEAD takes all these existing ideas about leadership, puts them together, and gives guidance around what, how, and when. LEAD is an *integrative* model. It connects the theory of safety leadership to the practice using a simple framework.

Think about a leader you admire. Ideally it will be someone you work with (they don't have to have formal leadership authority), but it can be absolutely anyone – a world leader, an actor, a politician (yikes!), or a sporting hero. Think about what she/he does to influence others to follow them and jot your thoughts down: What types of safety leadership skills, behaviours, or practices would you show if your team had made a mistake, error, or near-miss?

What does he/she do?

What does he/she say?

What might he/she think?

Now, consider how these characteristics might apply to safety. Does anything change? Probably not. Safety leadership is the practice of good or effective leadership in the domain of safety. By getting general leadership right, we can improve our safety leadership.

Now I would like you to consider how your safety leadership might change based on the type of situation you and your team are working within.

What types of safety leadership skills, behaviours, or practices would you show if your team was doing routine, everyday work, where work is low risk? Write down some of your ideas.

Maybe you wrote down things like: helping my team to be successful, encouraging effective work behaviours, assisting my team to be confident and capable, easing back on arduous safety requirements, getting my team the tools and gear they need to be successful, or keeping communication channels open.

What types of safety leadership skills, behaviours, or practices would you show if your team was pursuing opportunities (e.g., a new safety initiative was being introduced)?

Did you write down things like sharing decisions with my team, consulting with or asking my team for their input, creating a plan or vision about what the future looks like, allowing my team to think about how they can contribute, and helping my team to grow new skills?

What types of safety leadership skills, behaviours, or practices would you show if your team had made a mistake, error, or near-miss?

Perhaps you wrote down things like: encouraging people to speak up, asking my team to fix their errors or mistakes, and sharing and discussing what we learned so we can improve.

And finally, what types of safety leadership skills, behaviours, or practices would you show if your team was working in a high-risk environment (e.g., working at heights, working with suspended loads, working alone for extended periods)?

You could have written down things like: emphasising the rules or standards that apply, monitoring the team for problems, correcting mistakes or deviations from set ways of doing the job, holding team members to account, or making your safety expectations clear and widely known across the team.

See, you already know this stuff!

If you answered differently, that's OK. LEAD is not about prescribing a specific set of things you must do. It is about opening your mind to some potential opportunities to think about safety and leadership differently.

LEAD is about doing more of the stuff you already know, in the best ways possible, in times when it counts the most.

Finally, we'd like you to think about the different work situations that your team might experience. What are some examples of low-risk and routine work? What are some examples of opportunities to innovate or do new things with safety (the opportunities)? What are some examples of emergencies where you need to recover performance, or where a mistake, near-miss, or safety incident has already occurred? And lastly, what are some examples of high-risk work situations where we might be best served by reverting to our traditional methods of managing safety?

Write your ideas down.

Opportunities to do new things and innovate around safety	Emergencies, mistakes, errors, and safety incidents
Routine, everyday, and low-risk work	High-risk or dangerous work

Let's move onto the next idea.

Safety leadership is about resolving or reconciling important trade-offs. In the modern workplace, we are bombarded with a tonne of different work goals, some of which compete or conflict with each other. For example, take the fundamental 'production–protection'[2] trade-off (others have called this 'efficiency-thoroughness'[3] or 'production versus safety').

Now, we're not going to lie. We won't say that good leadership is about always coming down hard on the compliance side of the scales. In reality, safety waxes and wanes in importance; the situation changes, and different leadership approaches are needed to optimise a team's performance.

However, when the job is everyday, routine, and low risk (e.g., carrying a toolbox to the work site), we can afford to ease up on the safety prioritisation and instead focus on getting the job done successfully (and efficiently). Why write a safe work methods statement or procedure for carrying a toolbox? Yes, we've actually seen this in an organisation (remember Taylor from the previous chapter).

In the LEAD model, there are two fundamental trade-offs or dilemmas to resolve. The first is prevention versus promotion, which bears a lot of similarities to the production–protection dilemma raised above. On the one hand, prevention is needed when danger lurks close by: when we need to slow down, be cautious, careful, and vigilant; when security and safety are paramount.

On the other hand, promotion is needed when we want to be success-ful. When there are opportunities to be realised and creative solutions to be developed. Safety is equally about prevention (think high-risk work and recovering from accidents or mistakes) as it is about promotion (think 'pur-suing opportunities' and learning).

The second trade-off is flexibility versus stability. This trade-off has to do with how we manage uncertainty – that is, having incomplete informa-tion about our work or environment. Sometimes, we resolve this trade-off by reducing uncertainty. Drawing on procedures, processes, rules, and other strategies, we can make our work stable, predictable, and under direct control.

At other times, we resolve this trade-off by increasing uncertainty. Think about the last time you learned a new skill or stepped out to explore something new. In these situations, you were embracing uncertainty. For safety, we need a balance between flexibility (think innovation and continuous improvement) and stability (think routine, everyday work and dangerous situations where deviation from procedure could spell disaster).

We'll cover these trade-offs in more detail later.

The final idea of LEAD is the bundles of practices. These get to the heart of LEAD, as they are the practical things we can actually do as safety leaders to influence work outcomes. In rough terms, these practices help us to influence our workers to resolve the fundamental trade-offs in ways that maximise performance.[4]

Here are the four bundles of practices, in short detail.

Leverage encourages promotion and stability. Showing Leverage requires us to use practices such as reward and recognition for successful safety performance; setting goals and giving feedback as people progress towards their goals; high-quality coordination and communication within and between teams as work is carried out; and effective forward-planning practices that include the identification of safety goals.

Energise encourages promotion and flexibility. Energise is proactive, giving teams the capability to maintain safety in a dynamic and changing work environment. Energise exerts influence on safety through practices such as communicating a specific vision or direction for safety-related change, encouraging collective purpose and commitment to goals, and enabling autonomy and growth opportunities for team members. Energise also requires a consultative and participative approach to safety.

Adapt encourages flexibility and prevention. Adapt consists of practices such as emergency readiness routines, reflection on past performance, and error management. Emergency readiness routines remind workers of the dangers and risks they can face; continuous improvement practices such as After Action Reviews encourage critical reflection-on-practice and learning; and error management techniques such as talking openly about mistakes and fixing problems before they become major issues foster a preventative and flexible approach.

Defend encourages stability and prevention. Defend is best described as a way to reduce uncertainty that concentrates on stability and prevention. Drawing workers' attention to rules and standardised procedures centralises control and achieves reliable and stable operations. Defend requires leaders to emphasise standardised rules and procedures around high-risk work (e.g., 'Golden Rules', Safe Work Methods Statements – SWMS), highlighting legislated duties and obligations (e.g., safety acts and regulations), carefully monitoring compliance and initiating corrective actions when expectations aren't met, and highlighting sources of danger and risk in the work environment.

And voila, here we have it. The LEAD model.

Reflection Questions

- How mindful or aware are you currently of the team's operating situation? Could you name it if asked?
- How could you become more aware of the team's situation and build in some thinking time to consider how your leadership style should change to suit the situation identified?
- What's your understanding of the term 'trade-off'? Why are trade-offs important to identify, talk about, and resolve when safety is involved?
- What behaviours can trade-offs create if they are handled/resolved poorly? For example, think about the production–safety trade-off discussed briefly in this chapter.

Practice Points

- Personality and other individual factors will undoubtedly shape a leader's style, resulting in preferences for certain types of activities over others; however, effective leadership is learned over time and doesn't rely on having a specific 'personality for leadership'.
- The LEAD model combines many different theories and ideas about leadership into one overall framework, so by learning about LEAD, you will become a more balanced and capable leader generally.
- LEAD suggests that you will likely be performing many of the skills required to create high safety performance within a team; however, the key point is that LEAD helps you to consider the role of the situation in determining which practices to use when.
- The four LEAD dimensions – Leverage, Energise, Adapt, and Defend – resolve fundamental performance trade-offs in different ways; so be mindful of what compromises you are making when choosing specific LEAD practices (e.g., creating an environment rich in promotion and flexibility will encourage proactive and creative

behaviours, whereas emphasising prevention and stability will encourage greater rule-following and deference to established ways of doing things).

References

1. House, R. J., & Howell, J. M. (1992). Personality and charismatic leadership. *The Leadership Quarterly, 3*(2), 81–108.
2. Reason, J. (2000). Safety paradoxes and safety culture. *Injury Control and Safety Promotion, 7*(1), 3–14.
3. Hollnagel, E. (2017). *The ETTO Principle: Efficiency-Thoroughness Trade-Off: Why Things that Go Right Sometimes Go Wrong*. Boca Raton, FL: CRC Press.
4. Casey, T. W., Neal, A., & Griffin, M. (2019). LEAD operational safety: development and validation of a tool to measure safety control strategies. *Safety Science, 118*, 1–14.

3

Where Have We Been?
Where Are We Going?

Chapter Summary

Like many other concepts in safety and management science, everyone has their own ideas and definitions about what leadership is and isn't. This creates a lot of confusion, particularly when 'safety' is put in front to create 'safety leadership'. Walking through the academic literature fails to give us much clarity because there has been an overreliance on just one theory of leadership: transformational leadership. Leadership is a wide field, and when thinking about how leaders contribute to safety performance, an equally broad approach is needed. Researchers are only beginning to scratch the surface of what it means to be a safety leader, with the most recent studies indicating that all general leadership models apply and help us to understand what behaviours should be applied under what conditions and when. Transactional, transformational, authentic, and servant leadership are all models that we can draw on to develop a full repertoire of practices to improve safety performance. While debates and discussion still rage around general leadership versus safety-specific leadership, the LEAD model provides a way forward by assuming that safety leadership is good or effective general leadership practices, just applied to safety settings.

Much has been said and written about safety leadership. Entire conferences are devoted to the topic. Academics are feverishly researching it in the hope of identifying what makes safety leaders 'tick'. Yet, despite all of this, we still know very little about both the science and practice of safety leadership.

For example, there is still no consistency in how it should be defined. Researchers use all kinds of measurement tools. People are still unclear about what they should do to demonstrate safety leadership in the field. This makes it a tricky space to operate within as a leader out there doing the hard yards. But it also presents an opportunity to clean things up and move things forward.

LEAD brings together many different ideas about safety leadership and puts them into a package. Not only does LEAD integrate, it advances things forward.

LEAD introduces the idea that the mode of safety leadership needs to change based on the type of work that people are undertaking. As you will soon see, this is known as 'situational leadership'.

Most importantly, LEAD isn't made up of a heap of new practices that you need to understand and learn. You are probably doing many parts of the LEAD model already. The trick is to think about the 'why' and 'when'.

However, to use LEAD well, we need to know its heritage. We need to know where we have come from to know where we are going. Let's look at the history of safety leadership.

One of the key questions faced by safety leadership researchers is the divide between general leadership, safety-specific transformational leadership, and what I call 'vanilla' safety leadership. This debate is often overlooked or ignored by the academics, which means they tend to put safety leadership research into one large body of work. As a result, there is no firm position on whether safety leadership exists, let alone how it is best measured or improved. This threatens to undermine progress.

A Brief History of Safety Leadership

Safety leadership, as a distinct term, first appeared in the mid-1980s and early 1990s. At this time, conference publications by safety engineers and other industry professionals started to talk about safety leadership. Definitions of safety leadership were informed by observation and reflection, rather than detailed studies. For example, Carrillo and Simon[1] described safety leadership as a 'grass roots' style of leadership made up of practices like consultation, participation, and establishing safety as a core value.

Also in the 1980s, Dov Zohar[2] developed the concept of 'safety climate', which emphasised ways that managers drive safety performance, mainly by creating a shared social environment where safety was valued and prioritised. Management commitment to safety, a common dimension of safety climate,[3] has many similarities to safety leadership models and ideas.

In 1994, Simard and Marchand[4] started to explore safety leadership using a more rigorous approach. At the time, it was becoming very clear that leaders, and particularly frontline supervisors, played an important role in driving safety outcomes. Although 'safety leadership' and even 'leadership' weren't mentioned by the researchers, they showed that when supervisors visibly participated in safety activities, this drove positive safety performance among workers.

In 1999, two organisational psychologists, Hofmann and Morgeson,[5] explored how a theory called 'leader–member exchange' could explain the relationships between leadership, safety communication, and accidents. They proposed that when leaders have high-quality relationships with their teams, this created a desire among workers to pay back or reciprocate through their job performance. The study supported this idea in a safety context, with higher-quality leader–team relationships improving safety behaviour.

It wasn't until 2001 that O'Dea and Flin[6] brought safety leadership into the mainstream. They defined safety leadership as a participative style of management where leaders performed safety activities themselves (instead of relying on safety personnel) and had frequent, informal communications with workers about safety. Comments from workers in the research showed four themes or dimensions of safety leadership among offshore oil and gas managers: visibility, relationships, workforce involvement, and proactive management.

One of the most influential studies of safety leadership was done by Julian Barling and colleagues.[7] In this study, safety-specific *transformational* leadership, which is drawn from Bass and Avolio's full range leadership model, was the focus. The scholars defined safety-specific transformational safety leadership as: 'leaders who inspire, intellectually stimulate, and consider workers as individuals in the context of safety'. Barling was among the first researchers to argue that general leadership is different to safety-specific leadership.

At the same time, Dov Zohar (the safety climate guy) had branched out to explore leadership in safety. He approached the issue from both a general leadership perspective *and* a safety-specific leadership perspective. In his first study, Dov found that the effects of *general* leadership on safety climate and injuries were found to be affected by the leader's perceived safety priority.[8] In other words, general leadership improved safety performance only if there is a strong perceived commitment to and priority of safety by the leader. In the second study, workers were asked to think about how they communicated with their superior, and the responses were categorised into either safety-oriented or production-oriented, or both.[9] Training was designed to increase the frequency of safety-related communications between supervisors and workers, which was found to improve safety performance. Therefore, this second study provided some evidence that safety-specific leadership was a more important concept than general leadership.

After this period, research on both general leadership for safety and safety-specific leadership exploded. However, given the origins of safety-specific transformational leadership, researchers have started to go back to the drawing board to rebuild the concept of safety leadership from the ground up. In the next section, let's take a deeper dive into the general leadership and safety field.

General Leadership and Safety

The first approach to safety leadership is to apply existing theories of leadership directly to safety contexts without changing anything. The argument for using general leadership to explain safety performance is explained below.

The lived reality of leaders is that multiple goals and demands compete for their attention at the same time. Leadership is something leaders do constantly, not in separate silos of 'safety', 'productivity', 'quality', 'staff well-being', et cetera. If safety leadership is a specific concern, then general leadership researchers say we should consider whether leaders place priority on safety, rather than thinking about a fully separate leadership style or mode just for safety.

There are roughly five leadership theories that have been applied to explain safety performance: transformational, transactional, leader–member exchange, empowering leadership, and authentic leadership. From research, we know that leadership styles are related to important safety outcomes, ranging from safety climate to safety behaviours (typically compliance and participation), and also non-safety outcomes such as job satisfaction and commitment to the organisation.

General leadership is usually described in terms of language like 'styles'. The use of 'styles' is a problem, because it suggests that a style is hardwired or part of someone's personality, rather than something that can be nurtured, grown, and developed. We prefer the term 'modes' when it comes to talking about leadership generally, and safety leadership specifically.

Modes are like gears in a car. When we start off, we put it into first, rev up, and take off. Then the situation determines or influences our next step. If we need to slow down, we shift to a different 'mode'. Speed up, and it's a quick change into a higher gear. Need to park? Jump into reverse and do what needs to be done. The same can be said for leadership. LEAD takes this idea of gears and modes, and runs with it. We'll explore the different LEAD modes in later chapters.

Safety-Specific Transformational Leadership

One of the most popular approaches to thinking about safety leadership has been to use what is known as the 'full range leadership' model.[10] The full range leadership approach (typically shortened in scope to just include transformational leadership) has dominated safety leadership research and practice.

Transformational safety leadership comes from work by organisational psychologists, who changed existing and established measures of leadership such as the multifactor leadership questionnaire (MLQ) by adding the word 'safety' to the questions. Others applied parts of the full range leadership model such as laissez-faire leadership (passive or disinterested) and transactional leadership (focussing on minimum work performance like compliance) to safety.

One study done over a decade ago remains the most important because it broke new ground in this area of safety leadership. In 2002, Julian Barling and other researchers were the first to coin the term 'transformational safety leadership' and explored its relationship with safety climate, and a new variable they termed 'safety consciousness' (general awareness of safety issues at work).

Transformational safety leadership was linked with safety climate (shared perceptions of the value and importance of safety) and safety consciousness (paying attention to safety), which in turn was linked with safety injuries and incidents. In short, this study showed that safety leadership is an important part of reducing injuries in the workplace.

More recently, a study by Sharon Clarke in 2013[11] showed that transformational safety leadership was strongly related to proactive and citizenship-oriented safety behaviours (in other words, going above and beyond the job description for safety), whereas transactional safety leadership was related to safety compliance behaviours (following the rules).

This result painted a picture regarding the impact that frontline supervisors can have over worker behaviour through the use of different leadership styles. If you need to work on safety compliance, then use the transactional approach (applying rewards and punishments, correcting non-conformance). If you need people to be motivated and show proactive safety behaviours, be more 'transformational' or inspiring around safety.

If only it were that simple!

'Vanilla' Safety Leadership

Under this view, safety leadership is described as a distinct style or set of behaviours that are separate from general leadership.

Safety leadership is seen as either (1) a group of the most important general leadership practices as used by leaders in safety-specific settings or (2) particular safety-specific behaviours shown by leaders as they go about influencing others to work safely.

Let's look at (1) first.

Analysing in-depth interviews with a small sample of leaders at different levels of a mining organisation, Sarah Donovan and other researchers then 'coded' every leadership decision and behaviour against existing leadership models.[12]

A spread of safety-specific actions was identified against these models, ranging from 'displaying commitment to safety, maintaining a safe working environment' (transformational safety leadership) through to 'intervening when others don't follow safety procedures' (transactional leadership) and 'willing to admit mistakes' (authentic leadership).

Also, Sarah's study identified distinct patterns of safety leadership at different levels of the organisation, and external to the organisation (i.e., regulators), that all combine and influence safety performance at the 'sharp end'.

This study showed that we need to think about safety leadership as a broad concept that covers practices within all major leadership theories in the context of safety, as well as the ways that different levels of leadership interact and support (or not) each other's decisions.

So, rather than thinking about safety leadership as being separate to general leadership, Sarah's study instead points to general leadership practices that are used in situations where safety should be prioritised.

Let's now look at (2).

In 2015, Luke Daniel[13] used a bunch of interviews to investigate safety leadership in the Australian construction industry. After in-depth interviews with 20 leaders at different levels, Luke created a model of construction safety leadership.

Luke found that qualities shown by good safety leaders in construction included applying discipline, talking about a safety vision, showing honesty, and role-modelling good safety practices.

Summing Up and Moving Forward

General leadership and safety have received a lot of attention from researchers, but perhaps have been underused by industry in the safety setting.

Perhaps in the same ways that safety culture and safety climate created huge programs of research and lots of different ideas, safety leadership has also followed the same path – we can't see the forest for the trees. And that has meant a lot of 'junk' or dodgy ideas about safety leadership have been allowed to flourish. Also, it means that of the 'good' stuff, or high-quality research work, there is simply too much out there to make practical sense of it.

Anyway, lots of studies show that there are many different general leadership theories that predict safety performance. This is an area where LEAD shines – taking many different ideas about leadership and putting it into one package.

Safety-specific transformational leadership has dominated this area. According to this view, a good safety leader shows transactional and transformational safety-specific behaviours. But it still doesn't quite 'get there' in terms of providing a road map about what to apply when, and how.

The final type of safety leadership research, and where the least work has been done so far, is about attempting to identify the 'secret sauce' of safety leadership – specific behaviours that are unique and specific to safety.

However, attempts to identify exactly what ingredients are in this secret sauce are still going on today.

The LEAD model neatly builds on this work and instead treats safety leadership as general leadership behaviours applied in the workplace as people go about their different work activities, and trying to manage safety at the same time.

LEAD takes the best ideas from general leadership theory and applies them to the safety setting. These models include:

- Transactional leadership, where the focus is on driving compliance by correcting violations and recognising good safety performance,
- Transformational leadership, where the focus is on coming up with a future plan for safety, role-modelling, and relationships with workers,
- Authentic leadership, where the focus is on being humble and open to feedback, and sharing errors/mistakes with others, and
- Servant leadership, where the focus is on coaching and developing workers, and acting as an advocate for their needs and requirements around safety.

Where these ideas and models often fall short is in the application.

LEAD explains 'what and when to apply' safety leadership, namely, what we might call 'situational leadership'. Situational leadership proposes that successful leaders adapt their approach based on the requirements of the work situation.

In 2017, we applied these ideas about safety leadership to develop the LEAD model.[14] Specifically, leadership was described as four bundles of safety-relevant practices: Leverage, Energise, Adapt, and Defend. Each of these bundles was called a 'control strategy'. These strategies align with a specific work situation to achieve the best safety performance.

From this perspective, leadership is considered from the angle of its effects on the team rather than the behaviours and qualities of the leaders themselves. LEAD also provides a scaffold that links specific leadership practices to workers' performance via some concepts we will come to appreciate a little later in this book.

What should organisations and individuals do considering all this discussion and debate about safety leadership? Where should they turn when it comes to safety leadership in practice?

Here are some ideas.

Firstly, rather than narrowing our focus to safety-specific transformational leadership, we should invest our effort in general leadership training in the first instance, and secondly, try to understand the organisational culture to determine how much safety is prioritised and valued. Developing a culture in which leadership naturally supports pro-safety decisions and actions is likely to be more effective than just teaching people how to be 'safety leaders' alone.

Developing a safety culture starts with leadership. Getting the leadership part right by paying attention to safety, recognising safe performance, highlighting accountabilities for safety, crafting a plan (or dare I say it – a vision) around safety, all contribute to culture because, as the saying goes, 'what interests my leader fascinates me'. Stepping up to the plate with safety leadership is one sure-fire way to make a difference around safety and improve the working lives of those in your team and beyond.

Also, in the field, we should be clear about the boundaries between safety management and safety leadership. Focussing 'leadership' training and development on topics that include advanced communication skills like influencing, persuading, and supporting and focussing 'management' training on topics such as planning, risk assessment, and incident investigation will allow leaders to develop both technical and 'soft' skills that allow them to advance safety in organisations.

In short, don't confuse safety leadership with safety management. They are two separate concepts, equally important, but ultimately focussed on two separate goals: influencing others versus ensuring safety is monitored, measured, implemented, and achieved.

Reflection Questions

- What leadership theory do you think you overemphasise or show the most in your safety practices? What about the one you least emphasise or demonstrate?

- Why do you think this is the case? Consider the role of your organisation's culture, other leaders in your workplace, peer pressure/expectations, your team, and even personal factors like your personality.

- Think about what area of leadership most interests or excites you. How can you pursue development in that area, such as through a mentor or a peer coaching program, or even formal training if available in your organisation?

Practice Points

- Safety leadership is much broader than a single theory – transformational leadership is just one of many general leadership theories that have been applied to safety.
- Effective safety leaders not only are good general leaders, but show a genuine and meaningful commitment to safety (e.g., prioritising safety over production when it matters most), participate in safety activities, and engage with workers to hear their ideas and views – these are the most consistent 'safety-specific' leadership behaviours found through research.
- Studies have shown that different types of leadership practices encourage different safety behaviours, for example, focussing on rules and standards encourages compliance behaviours, and focussing on empowerment, consultation, and worker's involvement in decision encourages proactivity and going beyond the minimum standard (at least, in Western cultures).
- Some people have argued that safety leadership is actually a macro property of a team or an organisation that 'emerges' from the actions, inactions, and interactions of multiple leaders – these studies challenge us to think more broadly about safety leadership and how our individual actions can affect or influence others to produce an overall result.

References

1. Carrillo, R. A., & Simon, S. I. (1999, January). Leadership skills that shape and keep world-class safety cultures. In *ASSE Professional Development Conference and Exhibition*. American Society of Safety Engineers.
2. Zohar, D. (1980). Safety climate in industrial organizations: theoretical and applied implications. *Journal of Applied Psychology, 65*(1), 96.
3. Flin, R., Mearns, K., O'Connor, P., & Bryden, R. (2000). Measuring safety climate: identifying the common features. *Safety Science, 34*(1–3), 177–192.
4. Simard, M., & Marchand, A. (1994). The behaviour of first-line supervisors in accident prevention and effectiveness in occupational safety. *Safety Science, 17*(3), 169–185.
5. Hofmann, D. A., & Morgeson, F. P. (1999). Safety-related behavior as a social exchange: the role of perceived organizational support and leader–member exchange. *Journal of Applied Psychology, 84*(2), 286.
6. O'Dea, A., & Flin, R. (2001). Site managers and safety leadership in the offshore oil and gas industry. *Safety Science, 37*(1), 39–57.

7. Barling, J., Loughlin, C., & Kelloway, E. K. (2002). Development and test of a model linking safety-specific transformational leadership and occupational safety. *Journal of Applied Psychology, 87*(3), 488.

8. Zohar, D. (2002). The effects of leadership dimensions, safety climate, and assigned priorities on minor injuries in work groups. *Journal of Organizational Behavior, 23*(1), 75–92.

9. Zohar, D. (2002). Modifying supervisory practices to improve subunit safety: a leadership-based intervention model. *Journal of Applied Psychology, 87*(1), 156.

10. Bass, B. M. (1990). From transactional to transformational leadership: learning to share the vision. *Organizational Dynamics, 18*(3), 19–31.

11. Clarke, S. (2013). Safety leadership: a meta-analytic review of transformational and transactional leadership styles as antecedents of safety behaviours. *Journal of Occupational and Organizational Psychology, 86*(1), 22–49.

12. Donovan, S. L., Salmon, P. M., & Lenne, M. G. (2016). Leading with style: a literature review of the influence of safety leadership on performance and outcomes. *Theoretical Issues in Ergonomics Science, 17*(4), 423–442.

13. Daniel, L. (2015). Safety leadership defined within the Australian construction industry. *Construction Economics and Building, 15*(4), 1–15.

14. Casey, T., Griffin, M. A., Flatau Harrison, H., & Neal, A. (2017). Safety climate and culture: integrating psychological and systems perspectives. *Journal of Occupational Health Psychology, 22*(3), 341.

4

Matching Leadership to the Situation

Chapter Summary

Situational leadership is not a new idea. For decades, researchers have investigated whether matching leadership practices to the team requirements and/or situational demands matters when it comes to driving performance. Consider a high-risk work situation. The last thing a leader will want to do is hold a committee, engage in a lengthy discussion, and decide through collaboration on the best path forward. To do so would probably increase the danger even further. Instead, a situationally savvy leader would recognise that a quick and effective decision is needed, drawn on either their personal experience or that of a suitable team member, to respond and lead the team to safety. In other situations, such as when a new safety initiative is being discussed or designed, a more consultative and empowering approach is going to be the best approach. Giving people autonomy when it matters is going to lead to the best likelihood of uptake and engagement in the initiative. Crucial to success in a situationally driven model of leadership is a mindful awareness of the current situation, suspending judgement and assumption-generating, at least temporarily, to let the real world into your conscious awareness. Without mindfulness, a leader could miss important or subtle cues that signal what is really going on in the situation.

Take a moment to reflect on *when* you show certain leadership behaviours. It's highly likely that the demands of the situation influence or even determine the flavour of leadership that you show in any given moment. Even if you aren't aware of it.

Consider a situation where there is an urgent risk or problem to be dealt with such as signs of an electric fault in machinery being used – a situation where urgent action is required. You're hardly going to wait around for the hazard to be triggered, causing harm and loss to you or your team. When the situation requires it, leaders become directive, exerting their authority to make the decisions and take the actions that need to be taken to avoid catastrophe.

At other times, it is likely that you take a more measured approach. Say, for example, a team member is struggling with a safety problem. Perhaps you gather the team together to brainstorm, or source the resources he/she needs to be successful at solving the issue. When time is ample, a more participative and intellectually stimulating leadership style can be used.

The idea that a leader's behaviours or practices should be matched to the situation is not new. In fact, it has been around for a few decades. 'Situational leadership', as a theory, proposes that effective leaders develop an understanding of the situation and apply the most appropriate response.[1] This theory challenged other dominant ideas at the time, such as the role of a charismatic leader in rallying a team around a cause and drawing on their expertise to solve problems.

Situational leadership in its original form proposed that the 'situation' to be aligned with wasn't necessarily what was going on with the work, but instead the requirements and maturity of the team. A team that requires close monitoring and support to be successful will likewise require a more task-focussed (i.e., directive) leadership style. On the other hand, a team that is more independent and proactive will require a facilitative and supportive leadership style. Or so the theory goes.

Consider the case of a team that is made up of largely new employees. In this case, their uncertainty about how the job should be performed might best be met with a correspondingly directive leadership style. More experienced employees can instead have tasks delegated to them, with the leader instead focussing on relationships and providing support where needed.

Unfortunately, the theory of situational leadership is largely just that, a theory. The evidence is scant on the ground to support this approach in practice. Although the idea of using certain bundles or styles of leadership to meet the demands of the situation has some support, matching these behaviours to the maturity of followers unfortunately has much less.

One challenge for theories of situational leadership is figuring out what 'situations' were important for leaders to consider. The examples above included a situation where there was more danger and a situation where employees were less experienced. We can also identify other important 'situations'. For example, when a business faces intense competition from similar businesses, there might be increased pressure to reduce safety priorities. The experience of the leadership could be a critical situation. When leaders have less experience, then a more vigilant approach to safety could be needed.

This situation is becoming ever more important where technology is changing rapidly. Technological change can bring immediate safety improvements. For example, using robotic cameras for inspections keeps people out of hazardous places. However, the changes also mean people are doing different tasks and often learning new skills. This situation requires leaders to constantly think about how to change and adapt, keeping their team up to date with new skills and task requirements.

The LEAD model builds on the history of situational theories to take situational leadership in exciting new directions. Rather than matching the leadership style to the maturity of the followers, these practices are aligned with the task or types of work being undertaken, from a safety-specific perspective.

When a situation is high risk, a more directive and authoritative approach is needed. This is known as a 'transactional' approach that involves identifying and correcting issues before they become major problems.[2] It becomes quickly apparent that such a style would be essential for mitigating the effects of hazards, short-cuts, and other unsafe activities. Performing these leadership actions in practice requires leaders to constantly scan and attend to their environment, maintaining a strong vigilance over the work environment.

In a more routine environment, where risk is well-managed and the tasks are predictable, a leader again switches their key behaviours to optimise the team's performance. In such situations, the leader can concentrate on driving efficiency and achieving work goals, while reinforcing positive safety practices that support success. For instance, in such a situation, the leader acts to coordinate the team appropriately and encourage communication between team members. They also reinforce team practices such as cooperating, supporting, and other back-up behaviours.

When the situation has deteriorated and there is a real risk of problems occurring, or alternatively, things have gone wrong and the team is pausing to consider its next step, again a distinct set of behaviours is required – one that emphasises adaptation and flexibility, albeit within a framework that concentrates on prevention of harm and loss. In such situations, a leader may even call on his/her team to make important decisions, so errors are captured and prevented. Such a style also requires a leader to use practices that encourage an openness of opinion and speaking up behaviours among team members.

And finally, in situations where changes are being devised or implemented, such as new technologies, work processes, or systems, leaders must use an approach that emphasises the role of participative decision-making, whereby workers are given an active role in contributing to the nature of the change. This approach is essentially 'transformational' in that advanced influencing and persuasion skills are drawn on to capture attention and mobilise effort through tapping into internal motivation (or 'want to' motivation rather than 'have to').

This interaction of situation with leadership style gets to the core of the LEAD model. According to LEAD, an effective leader is 'mindful', in that they are acutely aware of the current situation. Not only are they aware, but they recognise opportunities to use the most appropriate leadership style for the situation. Take the following example of situational leadership in action:

One of the other areas I run for the company is where we do all the loading and pre-loading. It's a very, very busy area. It was a busy day, we probably had about 50 trucks pushed through on the day. The guys

> were all about what was in front of them. I got a feeling that we needed
> to stop, so everything's time slotted, everything's got to be out. With
> that mantra, I went out and stopped all my loaders, and we had a good
> clean toolbox on the spot, right in the middle of it all and it was about
> setting and resetting and regaining that focus. That we've got to stop,
> we've got to remember what we're doing and we've got to focus.

In this example, the leader deploys a Defend skillset in response to the
rapidly escalating work situation. The leader proactively 'senses' that risk is
increasing, as conveyed by weak signals (e.g., workers beginning to rush, a
long delivery truck queue, impatient drivers). As a result, the leader draws
on their Defend strategy, in particular, drawing attention to potential hazards and resetting workers' risk tolerance levels.

Another relevant example comes from healthcare. Researchers by the
name of Cook and Rasmussen[3] coined the term 'going solid' in relation to the
capacity of an emergency department. Going solid is a slang term borrowed
from nuclear power that means the boiler supplying turbines with steam fills
with liquid, dramatically changing the operating characteristics and risks
of the operations. In healthcare, going solid refers to a situation where the
capacity of a hospital to transfer and care for patients becomes more difficult
due to overloading. Essentially, there are too many patients and not enough
beds – the hospital becomes saturated with work.

When a hospital goes solid, it means that work operations become tightly
coupled, or in other words, closely dependent on each other, so much so that
small mistakes or problems in one area become magnified and affect other
areas (e.g., the decision to anaesthetise a patient in a busy hospital setting
means that other more needy patients are unable to access the care that they
need). How might a leader notice some of the signs of going solid? Increased
speed of decisions, pressure to discharge patients early, and faster pace during provision of patient care are all signals that a leader can be mindful
towards and proactively address through their LEAD strategies.

But what exactly does it mean to be 'mindful' anyway? Ellen Langer is
an influential thinker in this area.[4] Langer describes mindfulness as 'making novel distinctions'. But what exactly does she mean by that? In different
terms, making novel distinctions means paying attention to things that are
new or different in your environment. It is about noticing new things, and
being in a state of 'wakefulness' or being in the present moment. When we
are mindful, the following benefits are likely to occur:

- Higher sensitivity to the environment and work conditions
- Openness to new perspectives and information
- Development of new knowledge and perceptions

- Awareness of different perspectives (stepping into other people's shoes)
- Happiness, healthiness, and overall positive wellbeing

Mindfulness can even help to improve workplace safety. Mindful people are generally able to draw on greater levels of focussed and directed attention; they tend to notice more risks, make less biased or automatic judgements, and can consciously control their safety-related behaviours. Mindful people may also make fewer errors. Mindfulness has also been linked to improved situational awareness (put simply, 'knowing what is going on' and being able to anticipate what will happen in the future). Finally, mindful people are less likely to oversimplify problems and work activities, and think through the situation in more detail, rather than relying on automatic and fast decision-making (e.g., overlearned behaviours).

So how do we create a state of mindfulness and take advantage of these benefits? You may have heard of mindfulness before in other settings, such as in relation to Buddhism and meditation, but mindfulness is so powerful that it has made its way into mainstream psychology, healthcare, and even the business world. Sure, meditation is one way of reaching a mindful state. But it can also be done very simply by taking the time to notice what's going on around you, and searching for new information, and being open to different perspectives from other people (your manager, peers, and your team members).

A basic exercise that can help to appreciate the concept of mindfulness is as follows. Try it now.

1. Take your hands and overlap them, as if you are holding an open book in your hands.
2. Raise your hands slowly up and place them over your eyes.
3. View the world through the gaps in your fingers. Notice how much less you can see.
4. Take your hands slowly away from your face, noticing how much more information and richness becomes apparent about your environment.

This exercise is deceptively simple – it hides a great depth in how mindfulness can improve appreciation of the environment and the present moment. In the exercise, your hands are your thoughts, beliefs, biases, et cetera. Often, we walk around (metaphorically) with our hands over our eyes, taking in only what we want to see. Removing our hands and seeing the world in richer detail not only helps to improve safety, but also makes you a better leader.

The next opportunity you get, practise mindfulness. Use this mindful state in combination with the concept of situational leadership, which underpins the LEAD model. By becoming more mindful, you will be more aware of the situational requirements you face as a leader, and more rapidly discover which leadership strategy should be used to optimise the performance of your team.

Reflection Questions

- When, at work, do you find yourself 'zoning out' and being unaware of your surroundings? How does this affect your leadership performance?
- What are some ways you can remind yourself to practise being mindfully aware of your team's current operating conditions and situation?

Practice Points

- Practise being mindfully aware in your leadership role – noticing what your senses are telling you, suspending judgement just for a moment, and using this raw information to help decide what the best leadership approach is for the situation.
- According to the LEAD model, there are roughly four main situations that teams encounter: routine (low risk), high risk, change, and mistakes/errors/incidents.
- Notice what situation your team is experiencing in each moment, and connect the situation to one of the LEAD strategies.
- Use mindfulness as a way to stay tuned in to your team and to be aware of their current operating environment – it can be done in as little as a minute, yet carries powerful benefits if used regularly to the point of becoming a habit.

References

1. Blanchard, K. H., Zigarmi, D., & Nelson, R. B. (1993). Situational Leadership® after 25 years: a retrospective. *Journal of Leadership Studies, 1*(1), 21–36.
2. Avolio, B. J., Bass, B. M., & Jung, D. I. (1999). Re-examining the components of transformational and transactional leadership using the Multifactor Leadership. *Journal of Occupational and Organizational Psychology, 72*(4), 441–462.
3. Cook, R., & Rasmussen, J. (2005). "Going solid": a model of system dynamics and consequences for patient safety. *BMJ Quality & Safety, 14*(2), 130–134.
4. Langer, E. J., & Moldoveanu, M. (2000). The construct of mindfulness. *Journal of Social Issues, 56*(1), 1–9.

5

Safety as a Control Problem

Chapter Summary

When we say 'safety is a control problem', we mean that to achieve safety, the work process must remain in control. Control is an interesting word, with possibly some negative connotations (e.g., micromanaging, enforcing, punishing, constraining). Sure, sometimes we may need to exert control over a work process using top-down, effortful, and compliance-focussed strategies. However, at other times, work can paradoxically be controlled through a *lack of control*, at least in the traditional sense. Work can be controlled by giving people flexibility and autonomy to adjust, adapt, and improvise. For example, in situations where the procedure isn't up to scratch, would it better for people to blindly and mindlessly follow it? Or would it be better if they gathered together, conducted a risk assessment, planned out the task, and improvised? The key insight of this chapter is that control over work can be achieved by markedly different means, so a leader should be aware of how they are exerting control and use the most appropriate strategy in a given situation.

When people are asked to define safety, they tend to use the phrase 'reduction of harm to a level that is acceptable' or something similar. This is a fair definition, but what do we mean by harm? What is acceptable and what is unacceptable? Once you start scratching, major issues with safety definitions like this one start to appear.

Another way of defining safety is 'loss of control of a work process, resulting in the uncontrolled release of (potentially) harmful energies'. This definition suggests that safety is a control problem. Not in the strict sense that we hover over people and control everything that they do (this is a control strategy, but one that is very effortful and ultimately ineffective in practice), but rather in an abstract sense.

Safety as a control problem means that, to achieve safety, a leader must always remain behind the driver's wheel. That way, when conditions get slippery, or something unexpected happens, we stand the best chance of avoiding disaster.

Now, as we alluded to earlier, control can be achieved in a number of different ways. A safety scientist named Jens Rasmussen[1] eloquently described safety in the sense of control versus loss of control. He outlined a hierarchy of different control mechanisms, starting at the top with governments, who exert control on organisations through regulation and guidance (e.g., codes of practice). Organisations in turn exert influence over workers through instructions, rules, and procedures. Workers then exert control over work through performing safety activities.

Top-down control, although done by many organisations, is only partially effective. It suffers from a few serious problems. First, top-down control has a time delay. Instructions from the top take time to reach the bottom, which in a safety situation can spell disaster. Second, top-down control is effortful and resource-intensive. Third, top-down control is complex. To exert perfect top-down control, the layer above must be just as complex as the layer below. This means that there must be a procedure for every eventuality. It becomes totally impractical and even dangerous if procedures are followed too closely or don't quite fit the work situation. Just consider the oil rig accident on Piper Alpha – if employees had followed the standard procedures to gather in the crew quarters, they would have surely perished.

All in all, we should steer away from top-down control except where it is vital to have direct and tight influence over what is happening – for example, during high-risk work, where hazards are likely and carry a sizeable intensity. When work is high risk, centralised control can be helpful. Andrew Hopkins, a prominent safety scientist, describes this as an important determinant of safety in high-risk industries – rather than allowing for autonomy and discretion when dealing with highly hazardous energies, organisations should tighten their prescriptions for how work should be done. We take this a step further by proposing that leaders can implement more dynamic structures through the types of practices and behaviours that they show during high-risk work.

What if I told you that in other work situations (e.g., when implementing change, during routine work, and when mistakes are made) control could be achieved through relinquishing your grip over workers, in the traditional sense? It sounds counter-intuitive, or just plain wrong – especially when it comes to safety – but the LEAD model is based on this idea.

In situations where the unplanned or unexpected strikes, relying on a top-down method (i.e., waiting to be told what to do and when) is a likely way of experiencing problems. Take an aircraft carrier as an example. Aircraft carriers have been described as a 'high reliability organisation', which means they have a great potential for catastrophe, yet manage to avoid or absorb errors and other problems before they become major issues.

They are thought to be so successful because they have a raft of leadership- and culture-related characteristics. One such characteristic is 'deference to expertise'.[2] When you defer to expertise, you find the person(s) most qualified or experienced to tackle the specific issue, which might be the lowest in terms of formal rank or status. The newest member of the team might have the best idea on how to solve a safety challenge. In a top-down control approach, we waste that talent by deferring to hierarchy, waiting for the leader to 'make the call'. What if the leader isn't the best one in the room? What if there's no procedure for the situation? What if the 'new guy' has actually worked somewhere that dealt with this exact problem before and did so successfully?

In the LEAD model, Energise and Adapt create control over safety-critical processes through leadership styles that encourage ownership, autonomy, growth, reflection, and learning. They do this by practices such as consulting with your team, handing over safety decisions to direct reports, and encouraging an open and 'psychologically safe' team environment where anyone can speak up without fear of retribution or ridicule.

When uncertainty is high, or a change needs to be undertaken, the best way to exert control is through a bottom-up approach. From basic human psychology, we know that forcing change on others tends to result in bad outcomes most of the time. People end up working around the change, pretending to comply, or even actively resisting and undermining the change. We call this 'surface compliance',[3] which involves just going through the motions or ticking a box to complete a process with as little effort as possible. Surface compliance happens when we enforce one method of safety on every work situation, and rely exclusively on top-down control strategies.

Autonomy has long been realised as a crucial ingredient for human well-being, prosperity, and overall success. Take away someone's independence, and you take away a piece of what ultimately makes them human. Deci and Ryan,[4] two prominent psychologists, realised this fact about autonomy when they formulated their theory of motivation called 'self-determination theory' (SDT). SDT states that human motivation can be divided up into roughly three categories: 'don't want to', 'have to', and 'want to'.

In the 'don't want to' category, people are disinterested in the action or change. Their motivation prevents them from engaging in the task meaningfully. Studies have shown that some people are clearly in this category towards safety activities; they only do it when forced to or when punishment is applied.

'Have to' motivation means that the person realises that the task or action is required, but performs it to the minimum standard possible. Consider safety compliance – oftentimes people will follow the rules when a leader is watching, or when the rules state that it should be done. Such motivation is 'external' – it relies on someone or something outside the person to drive the activity. Not very sustainable over the long term.

'Want to' motivation exists when there is autonomy. Also important for this type of motivation is a sense of connectedness to others or a sense of belonging, and competence. When we give people independence (e.g., allow them to make their own decisions), give opportunities to talk and generally interact as a team, and grow, develop, and teach our people, they are more likely to be internally motivated. In other words, they do safety because they want to.

Recently, researchers have included notions of 'can do' in this bottom-up process. Having the confidence, skills, and belief that you can act on a problem comes when you are empowered to take control.

At other times, control should be achieved through traditional top-down methods because this is the most effective way of managing risk. When conditions are routine or work is high risk, usually we know with great certainty what is happening in our environment. Things are predictable and so we can increase our effectiveness (both in terms of efficiency and managing risk) through leadership activities that standardise, routinise, and generally treat work as a series of steps in a process. In the case of routine work, we find the best way(s) to do the task, document it, and encourage people to do more of that in ways that increase success. In the case of high-risk work, we apply the brakes and slow the job down, using our processes, procedures, and tools to take thorough stock of the situation in ways that prevent harm and loss.

In more detail, the LEAD model operates through inducing or encouraging a certain mindset among workers. This is because the practices we use form control strategies that frame work goals in different ways, resulting in corresponding mindsets or motivational approaches. Put simply, when we emphasise duty and obligations, this creates a 'prevention' mindset where workers concentrate on complying and minimising losses. In short, they will be much more conservative to risk. In contrast, when we emphasise growth and achievement, people are more likely to approach their goals using proactive and creative behaviours, or in other words, use a 'promotion' mindset. Let's explore these different mindsets in more detail.

Take Leverage to start with. This control strategy brings about a stability-focussed mindset through practices like clarifying goals and reinforcing established work behaviours that create successful outcomes (e.g., communicating with peers, supporting and backing up their team mates). When these practices are shown, workers understand that the goal is to create predictability in their work patterns. They also adopt a promotion mindset that encourages striving to maximise gains and successes. This is because recognition and reinforcement tap into motivational pathways that encourage an achievement orientation to work. So, work is likely to proceed more efficiently and effectively.

Energise has a slightly different effect on workers' mindset. Because Energise draws on practices like growth and inspiration, it too creates a promotion-focussed orientation where people are likely to be proactive and strive to achieve work goals. However, empowerment practices like participative decision-making and consulting with workers foster a flexibility mindset, which means workers are more likely to embrace change and explore opportunities to develop new capabilities.

Adapt changes the mindset over to prevention. Concentrating on mistakes and lessons learned sends a signal that prevention of losses is a priority. So, workers are more likely to approach the learning task by thinking of ways to be conservative towards risk. The mindset is also flexibility-oriented because learning concentrates on how to incrementally improve existing routines, processes, and systems rather than transformational change (as is the case for Energise). Engaging the team in 'what if' thinking that identifies where things can go wrong reinforces a prevention mindset but also encourages flexibility to develop new routines that reduce the scale and intensity of future failures.

Finally, Defend practices encourage a prevention and stability-focussed mindset. Highlighting sources of risk in the environment sends a signal to workers that they must be conscious of danger and taps into their needs for security and safety (avoidance of losses and harm). Monitoring performance and understanding the gaps between what is done and what is specified in procedures highlight the importance of deep and adaptive compliance – workers will be more likely to draw on standardised ways of doing things in a conscious and measured way to reduce risk. Fair and just accountability reinforces safety expectations and ensures that duties and obligations are embedded and made prominent for workers, again reinforcing their prevention orientation.

In all, safety is best treated as a control problem. The trick is recognising that 'control' means different things in different situations, so it requires special approaches from a leadership perspective. When work is high risk, we fall back to our traditional hierarchical control strategies. We monitor compliance and drive ownership and accountability through highlighting expectations and obligations. When work is driven by changing initiatives or new technologies, for example, or when mistakes are made, we need to utilise bottom-up approaches that leverage the expertise and experiences of our workers. We fully embrace their capabilities and encourage them to contribute to the solution.

In the next series of chapters, we will delve deeper into the LEAD model and explore specific strategies and practices you can utilise to drive exceptional work (and safety) performance.

Reflection Questions

- In what ways might the work your team performs get 'out of control' or otherwise increase in risk levels?
- What methods of control do you default to using to regain control in these situations?
- Do you tend to emphasise top-down or bottom-up strategies to exert control? Perhaps there are opportunities to practise other types of strategies that still get you and your team to its desired outcomes but use a different method to exert control.

Practice Points

- Consider how you are creating and maintaining control over work processes, and how the presence or absence of control affects your team's safety performance.
- Using different methods of control to influence work and safety performance results in the development of internalised or 'want to' motivation – people start adopting your goals as their own and self-starting their work activities without you being around.
- Each of the four LEAD strategies exerts control over workers in either top-down or bottom-up ways, or a combination of both; therefore, be mindful of what LEAD strategy you are using to achieve control and whether it is the best one for the situation.

References

1. Rasmussen, J. (1997). Risk management in a dynamic society: a modelling problem. *Safety Science, 27*(2–3), 183–213.0
2. Weick, K. E., & Sutcliffe, K. M. (2001). *Managing the Unexpected* (Vol. 9). San Francisco, CA: Jossey-Bass.
3. Hu, X., Yeo, G., & Griffin, M. (2018, July). Safety compliance re-examined: differentiating deep compliance from surface compliance. In *Academy of Management Proceedings* (Vol. 2018, No. 1, p. 15338). Briarcliff Manor, NY: Academy of Management.
4. Deci, E. L., & Ryan, R. M. (2008). Self-determination theory: a macrotheory of human motivation, development, and health. *Canadian Psychology, 49*(3), 182.

6

Leverage – Getting Things Done

Chapter Summary

Leverage is the safety leadership strategy that is used when the situation is routine and low risk – in other words, business as usual. Leverage concentrates on setting effective goals for safety (both personal leadership goals and team goals). Setting goals brings clarity and drives positive motivation towards their achievement. Clarifying roles and responsibilities is also a key feature of Leverage, ensuring that workers understand who has specific safety duties and what is required. Often we can make assumptions about safety, leaving too much open to interpretation and the potential for misunderstanding. A skill that goes hand in hand with setting goals and responsibilities is performance feedback, both positive and constructive. When goals are achieved, positive recognition reinforces safe ways of performing work, making them more likely to be shown again. Delivering constructive feedback effectively makes sure the message is received and acted on, and not just allowed to wash over and be ignored. Finally, a great leader also coordinates their team. Coordination doesn't just happen by accident or magic, a leader and their team must work hard to keep things together and on an aligned path. We explore strategies for coordinating a team in this chapter.

We now take a closer look at the core LEAD behaviours. Starting with 'Leverage', we explore the foundations of good safety leadership. Leverage is our starting point because it involves two basic principles of good management in any workplace: clarifying goals and providing feedback. People become frustrated if they're not sure what they should be doing and how well they are going. Over time, this frustration can lead to poor health and high levels of dissatisfaction and burnout. It is the same with safety; goals and feedback are essential.

Leverage is a good starting point because it gets things moving in your safety leadership plan. The word 'Leverage' draws on the idea that large changes can be made with some simple yet powerful actions that link our leadership actions to better safety outcomes. Clear and meaningful goals together with effective feedback provide the leverage for this change. Safety efforts begin when people have a clear idea of the safety goals everyone is striving for and how the goals benefit the whole workplace.

Clear goals help people know what tasks to do and why they need to be done. For example, a clear goal might be to ensure that every person working on isolated equipment should fit their own lock or danger tag to the piece of equipment. This is a clear goal that says 'what' should be done when isolating equipment. It is also important to link the goal to its purpose or 'why' the goal is necessary. Lock and tag-outs are generally understood to protect workers from unintended operation of hazardous equipment, but this purpose can be explained in terms of your own experience and workplace characteristics. Has someone been injured in the past because a piece of equipment was not locked? Are the potential consequences catastrophic for people and their families? Does following this procedure help others in the workplace? When people have a clear picture of a goal's meaning and purpose, they are able to think more carefully and accurately about how to do their current work.

When we set up clear goals with a purpose, people are more naturally motivated to strive towards the goal and figure out ways to achieve the goal. This last point is worth thinking about. People with clear and meaningful goals try to understand 'how' to achieve their goals. Stop for a moment and think about the different ways people might support the goal to have everyone fitting their own lock and danger tag. Some ideas include raising the topic at toolbox meetings, messages in daily briefings, signage on site. You will have more ideas, and it is important that goals leave room for people's ideas.

Clear goals motivate people to work harder and also work smarter. You do not have to specify every single step needed to meet each goal. Too much unnecessary detail can actually drain motivation. Most of the time we spend our efforts performing routine and everyday work. For instance, setting up for a simple maintenance job, repairing a leaky valve in a low-pressure process, or conducting a basic job site inspection. Complicated goals are not needed for these situations and can actually prevent people from figuring out the best way 'how' to do the tasks. The problem is that in some organisations, these low-risk situations are treated as though people's lives depended on them being safe, when in reality, the risk is incredibly low. Not much can go wrong, and we are well-familiar with the hazards and controls required to carry out such work.

In one organisation, safety had become so bureaucratised that a safe work methods statement (a prescriptive step-by-step safety procedure designed for high-risk construction work) was created for employees to carry a toolbox from their vehicle to the job site. Completely over-the-top and unrequired. Furthermore, all this did was undermine the culture of safety across the organisation. How could safety be seen as credible when such grossly disproportionate risk management strategies were being implemented?

The Leverage strategy concentrates on giving workers clarity of goals and roles (what needs to be done by who), reinforcing successful and effective practices (embedding good work habits), and coordinating the job through fostering positive and ongoing communication (developing a sensitivity to operations and shared understandings of the current state).

A lot of research has been done to understand what makes goals effective. For starters, ambiguity or confusion about a task or role is a source of psychological stress and even strain over the long term. When people experience too much uncertainty, particularly when they lack confidence or are under pressure, this can be harmful psychologically and shows in their lowered task performance. Too much certainty and prescription, however, and a leader runs the risk of eroding people's sense of self-determinism and autonomy. What is needed is the right balance between the two.

Two psychologists, Locke and Latham,[1] have extensively studied the benefits of bringing team clarity through setting challenging yet achievable goals within teams. Their program of research has spanned many decades and has found that certain types of goals can increase performance by as much as 250%. Goals need to stretch an individual's abilities and capabilities to have this dramatic effect on performance. Only when ability has been exhausted does performance start to drop. In practice, this means setting very specific and detailed goals, rather than general or vague goals. Consider this example, what do you find more motivating and challenging: 'Do your best' or 'Increase the quality of your safety incident reporting through concentrating on describing the environment or context in more detail'.

When a specific goal is developed for someone, a few things happen psychologically. First, they feel more energetic or motivated to achieve it. They also activate their existing knowledge and skills to direct towards the goal. And lastly, additional effort and persistence are invested. So, there are many advantages to setting challenging and specific goals with your team.

Consider the current state of safety in your team. What are some areas that the team could improve or direct their attention towards in ways that would stretch or challenge them?

Draft some SMART goals for your team around health and safety. A SMART goal is specific, measurable, achievable, relevant, and time bound (it has a due date or expiry date).

Consider sharing these goals with your team at your next meeting and negotiating them based on any feedback you receive.

A follow-on from goal setting is performance feedback which lets people know how well they are progressing towards goals. A common trap in safety leadership is to provide negative feedback about how people are 'not' reaching goals. For example, individuals not carrying out procedures correctly, or organisations not meeting injury targets. It is important to understand these negative outcomes, but it is even more important to provide positive feedback that paints a picture of how people are contributing to successful goals.

Positive consequences are *very* important for motivating actions at work. We do not mean that safety leaders must find a new set of rewards and consequences for workers who behave safely. We simply say that leaders who are Leveraging safety can communicate that safety behaviour has positive consequences, not just negative ones. The role of positive and negative feedback has been studied extensively over many decades. Albert Bandura is a psychologist who founded a theory called Social Cognitive Theory.[2] In this theory of learning, Bandura proposed that people learn through observing other people, ideally experts, and receiving performance feedback. Positive feedback can be particularly motivating because it taps into the neuroscience of reward. When positive feedback is received, chemicals such as dopamine are released, which encourage people to persist with the task and invest extra effort to succeed. Positive feedback reinforces existing pathways in the brain and can help to form new ones, such as learning a new skill or practice.

Feedback also builds something called 'self-efficacy'. Closely related to confidence and self-esteem, self-efficacy is the beliefs we have about our ability to persist despite experiencing setbacks and challenges. When self-efficacy is high, we are more likely to keep at a task, even when the going gets tough.

Many studies have suggested that workers are least satisfied with the ability of their supervisor to give adequate recognition and positive feedback.[3] Not only do workers want more feedback; they also want higher quality feedback. In fact, when recognition is done, we usually default to a simple 'well done' or 'good job'. When we do so, we miss an opportunity to consolidate learning and reinforce effective practice. Take a moment to note down your answers to the following questions:

Think about the last time someone gave you really effective feedback. What about the feedback was helpful for you? How was it delivered? What stuck with you and why?

Delivering good feedback isn't rocket science. In fact, it can be done with three simple steps.

1. Describe the situation.
2. Describe the actions (what he/she did).
3. Describe the outcome and acknowledgement.

We can call it the 'SAO' recognition model.

Consider an event or achievement that you could recognise within your team. Use the SAO model to write how you might approach that recognition task.

What was the situation?

What action or actions were performed by the team?

What were the outcomes (what was achieved)?

Coordination is a hallmark of an effective team. When a team coordinates their activities, they are dramatically more effective than if they were to operate independently. However, effectively coordinating a team is no easy feat.

In routine and well-known environments, teams draw on a number of important processes to stay informed and in control of the work situation. A group of psychologists (Marks, Mathieu, and Zaccaro) devoted significant time to studying team coordination processes and discovered what successful teams actually do.[4]

Successful teams use a range of different planning, coordinating, and general management processes, which leaders within a team can encourage and support. For example:

- Interpreting the mission/job: Identifying the main tasks to be done, the environmental conditions, and the resources that are available.
- Specifying goals: Forming specific goals for the team to strive towards, in this case, including safety goals as well as quality, innovation, efficiency, etc.
- Strategy: Developing multiple plans and pathways to accomplish the job, including anticipation of problems.
- Progress tracking: Checking whether the team is moving in the right direction towards the set goals, and communicating such progress information to team members.
- Conditions tracking: Checking whether resources or environmental conditions have changed, and updating the team if they have.
- Team monitoring and back-up: Encouraging team members to help each other complete tasks.
- Conflict management: Stepping in and helping team members to compromise and resolve disagreements/arguments when required.
- Motivation and confidence-building: Saying positive and motivating things that build the team's confidence, such as talking about examples of where they have been successful in the past.

Others have studied what makes for a successful team in environments like healthcare, taking research from high reliability organisations (e.g., nuclear power and military) to help them understand what these planning and coordination ingredients look like. Wilson, Salas, and colleagues found that 'high reliability teams' practise a number of special processes that boost their performance, including safety.[5] These processes include:

- Closing the loop: Team members repeat back what was heard and ensure the handover of information is accurate and complete.

- Clarity: Communicating in ways that ensure information is unambiguous and concise.
- Backing up: Stepping in to help each other when people show signs that they are struggling.
- Performance monitoring: Team members are aware of opportunities to give each other constructive feedback to improve performance.
- Shared understanding: Team members know each other's strengths, weaknesses, and details of what they each contribute to the job.
- Situational awareness: The ability of the team to share information as work progresses, and keep each other informed, as well as anticipating where things might change or develop into problems.

An area that deserves special attention is what is called 'shared mental models'. A mental model is basically a mental picture of how we expect the world to work (or more technically a map of psychological knowledge structures). When we talk about 'team mental models' we extend this concept from the individual to the group. A team mental model exists when a group of people working together has a similar understanding of the situation and how things should happen. On the plus side, a shared mental model like this helps cohesion, cooperation, and recognition of problems. For example, team members might recognise when a co-worker has forgotten a procedure and give a reminder or carry out the needed task themselves.

On the down side, mental models can act like blinkers, hiding bits of information that are different to our worldview or challenge existing beliefs about reality. Mining tragedies have occurred when people have assumed someone must have already carried out necessary inspections. These types of assumption are captured in the idea of team mental models. Safety leaders seek to unpack the assumptions of their team and develop shared mental models that will be effective.

Developing shared mental models in a team helps with safety, and overall performance, because everyone has a common understanding of the task or job to be done. They can more readily anticipate where problems might occur, where certain resources might be needed, or when risk is likely to increase. By conducting thorough team briefings and pre-starts, cross-training, job rotation, buddying, training, and shadowing, teams can develop very detailed and shared mental models. Their planning and performance will likely benefit greatly.

But what happens when the situation becomes novel or unique? How can a team remain successful in these conditions, with a leader's support? What typically happens in these conditions is that a team switches to 'implicit coordination'. Alternatively stated, the team saves its resources (to compensate for the increased demands of the situation) by talking less and relying on their shared understanding to coordinate work.

Airline pilots are a great example. During periods of low intensity, they will watch and observe each other in the cockpit, developing a good working knowledge and shared understanding of each other. When operating demands increase (i.e., they are under stress), their performance will be related directly to the quality and accuracy of their shared mental models. In other words, if a team member is unaware of what their colleagues do, the contributions and skills they bring to the table, and generally how they work together, performance will degrade when the heat is on.

Importantly, teams can be trained and encouraged to become better at dealing with unique situations. Such training is referred to as 'adaptive coordination training'.[6] Examples of this type of training include the following:

- Cross-training: Training team members to become familiar with each other's roles and responsibilities (e.g., job sharing).
- Overlearning procedures: Developing automatic and effective responses to situations that are likely to be encountered in the work environment (e.g., rehearsing a procedure multiple times).
- Introducing variability: Making a team's task more difficult or varied by making small disruptions or changes (e.g., limiting or rerouting communication channels that forces the team to compensate).

The last strategy, introducing variability, can be very effective at improving a team's performance under uncertain and dynamic conditions. By forcing a team to cope with variations, like disruptions, setbacks, or challenges, the team learns more effective and 'deeper' strategies to achieve success.

Reflection Questions

- How do you bring clarity to your role as a leader driving safety? In other words, how do you find out what you should be focussing on to improve safety?
- What is your organisation's current safety vision or strategy? Do you find any challenges translating that vision or strategy into tangible goals for your team to achieve?
- What is one thing you can concentrate on recognising among your team to improve their safety performance? Think about where the team has started the process of change with safety (e.g., adopting a different behaviour) and how you can accelerate this development through feedback.
- How actively involved are you in helping to coordinate your team? Are they fairly autonomous and independent, or do they rely on your guidance most of the time? Do you think you have the right balance of involvement in managing your team's coordination?

Practice Points

- Set SMART goals for yourself and your team that concentrate on safety, remembering that a SMART goal is specific, measurable, achievable, relevant, and timely.
- Take into account your organisation's overall safety targets or vision when setting team goals; for instance, what would your team be doing to help achieve the overall goal?
- Make sure the team members are clear about their roles and responsibilities for safety through team discussion and negotiation – rotate roles around the team to develop multiple skill sets and avoid redundancy, and where possible, allow team members to have some say over the roles in which they wish to grow and develop.
- Recognition for safety is one of the least frequent behaviours shown by leaders, so consider how you can incorporate meaningful verbal feedback into your daily team discussions.
- There are many practices that help to coordinate a team, but above all else, keep yourself in the field and stay connected to your team as much as possible.

References

1. Locke, E. A., & Latham, G. P. (2002). Building a practically useful theory of goal setting and task motivation: a 35-year odyssey. *American Psychologist, 57*(9), 705.
2. Bandura, A. (2001). Social cognitive theory: an agentic perspective. *Annual Review of Psychology, 52*(1), 1–26.
3. Kelloway, E. K., Nielsen, K., & Dimoff, J. K. (Eds.). (2017). *Leading to Occupational Health and Safety: How Leadership Behaviours Impact Organizational Safety and Well-Being.* New Jersey: John Wiley & Sons.
4. Marks, M. A., Mathieu, J. E., & Zaccaro, S. J. (2001). A temporally based framework and taxonomy of team processes. *Academy of Management Review, 26*(3), 356–376.
5. Wilson, K. A., Burke, C. S., Priest, H. A., & Salas, E. (2005). Promoting health care safety through training high reliability teams. *BMJ Quality & Safety, 14*(4), 303–309.
6. Gorman, J. C., Cooke, N. J., & Amazeen, P. G. (2010). Training adaptive teams. *Human Factors, 52*(2), 295–307.

7

Energise – Pursuing Opportunities

Chapter Summary

'Energise' is used when a team is undergoing some form of safety-related change, such as an innovation or initiative. The core idea with Energise is to adopt a bottom-up and inspiring approach that engages workers in the process of change. When this approach is used, change is more likely because people take greater ownership. Of course, people in your team won't feel equally comfortable with high autonomy and involvement. You will need to use your leadership skills appropriately, combined with your knowledge of team members' preferences, to use Energise effectively. Overall the best strategy is to create 'freedom within a frame' – defining general goals and objectives, building the team's confidence through positive talk like inspiration and examples of success in the past, and giving them the support and resources they need to initiate change. The last point is critical to achieve your objectives. A leader using Energise will work with the team members to build their capability, drawing on techniques and tools like informal training, coaching, mentoring, and buddy systems.

From time to time, changes must be implemented, not only in the way safety is practised, but also how work is done and what is used to perform work (e.g., new technologies). In modern workplaces, change is becoming even more frequent as companies struggle to compete and keep up with the unrelenting pressures of a global market and constant innovation. We hear words like 'disruption' and 'uncertainty' more often than ever. So, leaders need techniques and skills to deal with such change, particularly around safety.

For an organisation to be safe, it must develop what is known as 'dynamic safety capability' – a process of scanning for threats and opportunities, reflecting on what processes and activities exist currently, and identifying methods to change and improve.[1] Organisations must carry out regular safety change, both incremental and transformational, reactive and proactive. In short, to stay ahead of the safety curve and to keep improving, safety change must be continual.

When it comes to the LEAD model, Energise is the strategy that is used when change is being implemented or experienced. Energise draws its strength from a number of psychological ideas and theories that support a participative and bottom-up strategy to designing and performing change in organisations and teams. Mostly, people find change to be unsettling, and in some cases, quite threatening. Change can identify weaknesses, trigger anxieties, and stimulate concerns about one's ability to cope with increased demands and expectations that the change could bring. Rather than forcing or implementing change, which only results in resistance and push back in most situations, leaders can instead adopt a more empowering and what is known as an 'internally motivated' acceptance of the organisation's goals.

Energise concentrates on leadership techniques that are proven to help people not only to accept change but to actively participate in making positive change. Participatory styles of leadership, where workers are actively involved in decision-making, and consulted meaningfully, lead to better change outcomes.[2] Consultation before, during, and after safety-related change is also required by many countries' safety legislation. Consultation helps leaders and other major decision-makers like business owners or general managers to appreciate how the effects of their decisions could impact on other people's safety. This knowledge of the effects of decisions (like allocating resources, introducing new ways of working or new technologies) improves safety by considering how change could introduce new hazards or penetrate/render ineffective existing controls.

Yet many organisations fail to uphold and fulfil their consultation obligations beyond a simple one-way transfer of safety information (e.g., a leader describing what works will be undertaken and asking for questions or comments). Trust is a critical enabler of consultation – without it, a leader may ask for input and feedback yet get very little in return. To build trust, a safety leader must demonstrate their ability (they know what they are doing and admit when they don't), integrity (follow through on promises), and benevolence (show good intentions towards others).[3] When trust is formed, consultation will be richer and more fruitful.

Empowering workers by giving them opportunities to make decisions doesn't have to be rocket science. Take the example of introducing a new protective device. One strategy could be for the safety team to make an 'expert' decision on what is best and implement that top-down through the leader. Compliance may be a problem, as the gear likely won't be fit for purpose (e.g., safety glasses that fog up and actually introduce a hazard), so people work around it. Another strategy could be to give workers a few options to experiment with and allow them to make a purchasing decision themselves. Tapping into the fundamental human need for autonomy and self-direction means that leaders can make it more likely that workers will wholeheartedly embrace change and uncertainty rather than fight against it. The more control an employee feels he/she has over a change, the more likely they are to accept it.[4]

Think about some of the safety or general work decisions that you typically make on behalf of or 'for' workers. What is one decision about safety where you could empower workers to at least participate in the decision, or otherwise make their own decision about how the practice or change should be done?

When a team member is empowered, they have confidence, skill, and permission to do meaningful work. Your inspiration kick starts this process because inspiration helps make the job meaningful (safety matters!) and builds confidence (we can do it!). But to really empower someone, they need permission.

Why are we saying that people need permission to be empowered? Sounds like a contradiction! And in some ways it is. Because to lead safety you also need to let people lead themselves. This idea is often described in different ways. Researchers use terms such as 'autonomy', 'self-direction', 'initiative', and even 'self-leadership'. The key idea is that people will think for themselves and you will give permission to do that.

Of course, it's more than simply saying to your team 'you have permission to think for yourself'. But you might be surprised how often we see leaders give a subtle message that 'thinking for yourself' is discouraged and not allowed. And it's understandable. When jobs are dangerous and you feel responsible for keeping people safe, it can be the right thing to have a list of 'don'ts'. For example, always wear gloves, don't leave without them. It's reasonable and helpful.

One of us worked with a company who had greatly improved the use of gloves in a business unit producing chemicals. The change was part of a participatory process where workers were involved in planning and identified that wearing gloves was one of many ways they could improve safety. The manager of another business unit saw a safety report that described the much improved use of gloves. Based on this information, the manager sent a directive to all employees that they should immediately wear gloves more often. This manager was frustrated that the directive did not seem to result in any change. But this result was not surprising, given there was no empowerment at all in the process used by the second manager.

Another practice within the Energise strategy is inspiration. Derived from a stream of general leadership work called 'transformational leadership', showing inspiration isn't as hard as it might sound. As a safety leader, you can be inspirational. That doesn't mean you must be an extraordinary or exceptional leader. Not at all. In fact, you will best inspire others by simply being yourself and sharing how you feel. Yes, your own experiences and feelings are the key to inspiring safety. You will build a personal connection with your team when you can share something about yourself.

Talking positively and in an inspiring way generates momentum and energy to tackle a problem or change head-on. It provides reassurance that the team can achieve its goals. It builds confidence that the team can perform well and will be successful. Inspiring can be as simple as describing examples of where the team has achieved success in the past and emphasising that success during a meeting or gathering. A leader is inspiring when he/she just gives reassurance that the team is capable of changing and doing well. Not only can they draw on past examples, but they can also describe positive qualities and capabilities of the team that show success is possible.

People often respond to the 'vision' of inspirational leaders. A vision sounds like a grand plan, but it is really much simpler than that. A vision is a positive view of the future together with the belief that the team can make that future a reality. Your vision for safety and your genuine feelings about the vision are a force to get people motivated towards safety and believing they can make a safer workplace for everyone. You can really motivate people to think about their safety future. Your vision is also a step on the path towards empowerment.

There are two basic steps to the inspirational part of energising safety leadership. First, getting in touch with your own response to safety. Second, using that feeling and experience to build a picture of a future for everyone. This positive picture you paint is the much talked about 'vision'. It's simple but important.

Let's start with an activity around the first step.

Activity: When did a safety issue create a strong feeling for you? For example, you saw a serious near-miss and suddenly felt both relief for what didn't happen and fear for what could have happened. Hold that feeling and imagine how you would describe the experience to your team. Describe the feelings and thoughts that ran through your mind.

When you feel this response to safety, others will too, and that is an important start for energising the team. The next step is to direct that feeling. That's where a vision comes in. A vision gives a positive view of what the future might look like. The vision doesn't need to be grand or extreme; it just looks to the future with a positive feeling connected to your own personal response.

A vision can be as simple as 'we will be a team that works well together', or 'by the end of the project we will look back with pride on what we have achieved'.

Activity: Using the feelings generated in the last exercise, create some 'future' statements for the team: 'We can be a team ...', 'In this project we will ...'. Notice that you can make some fairly simple statements, yet appreciate these can be powerful, even inspirational, motivators for your team.

Remember we said previously that 'charismatic' leaders can be detrimental? The negative aspect of charisma can appear when these two steps are extreme or distorted. For example, if the feelings conveyed are not personal and the future picture you paint is not necessarily for the benefit of the team, charisma can hurt the team.

Our steps towards inspiration will help you to set the team on a genuine and meaningful path. You have now set the groundwork for empowerment, the next part of energising safety leadership, and we will explore this next.

Think about some of the challenges and adversities that your team or colleagues have worked through. What is the one example that you can remember and could use the next time you have to implement or support a particularly challenging organisational change?

Another practice within Energise is 'growth'. When a change is about to be undertaken, people will be concerned about their skills and abilities to perform it well. This concern could be a major barrier to change implementation. Models of change often include the concept of confidence or what is technically known as 'self-efficacy' – beliefs about your ability to persist and be successful under challenging conditions. Helping employees to 'grow' new skills helps to build their confidence.

Importantly, we learn by observing experts, making mistakes, and receiving constructive feedback on our performance. Experiential learning is best when teaching new skills or practices, as it gives a realistic example of how the skill can be used. Also, when people make mistakes or errors in natural or realistic environments, they experience a deeper level of learning than if they just rehearse or practice the perfect sequence of steps required to perform the job. When people make errors, and get feedback, they troubleshoot and develop rules and methods to cope with setbacks in the 'real world'.

A useful mnemonic that can guide the teaching of skills is 'MEDICS'. MEDICS stands for:

- Motivate: Provide inspirational language and explain 'why' something needs to change or be done.
- Educate: Provide detailed explanation of the change or practice.
- Demonstrate: Show an expert performing the behaviour or practice.
- Imitate: Ask the worker to show you or repeat back.
- Consolidate: Provide feedback on performance.
- Support: Take an ongoing interest in the workers' performance, offer feedback, and reinforce best practice.

A leader can also support growth by coaching. Coaching is a process of providing advice and feedback to help grow new skills or refine and master existing skills. A good leader is a good coach because they serve their team members and concentrate on equipping them with skills and knowledge that they need to be successful. When dealing with change and growth, a helpful model to consider is 'appreciative inquiry' or the 4D model: Discover, Dream, Design, Do.[5]

Appreciative inquiry is a process of working to people's strengths, helping them to imagine a positive future, and co-designing the solution to ensure personal relevance and overall successful achievement of change. Discover involves asking questions to uncover a person's strengths and weaknesses. In essence, you are interviewing them to develop a solid understanding of their capabilities. Dream is facilitating a positive future that is rich and descriptive. For example: Imagine it is 12 months into the future. What will you have achieved? What will be your successes and why? Design is about working with the person to set goals and identify a path forward. Do involves providing ongoing support and help to the person so they achieve the desired outcome.

In the next section, we invite you to conduct an appreciative inquiry process on yourself using the following questions as a guide. Use these questions to develop your own coaching process when working with the members of your team. Safety can be injected into this set of questions to focus the topic even further.

Step 1: Establish the Topic

- What are the times when you are at your best (safest) at work?
- Who do you respect for their work skills and abilities? Why?
- What would you like to do more of to improve your (safety) leadership?

Step 2: Create a Positive Vision of the Future

What does it look like when you are performing at your peak at work?

Fast forward 12 months. What would you have liked to achieve at work (with respect to safety)?

What does great (safety) leadership look like to you?

Step 3: Designing the Path Forward

What could you experiment with or try out at work?

What help do you need to change?

How could you gain feedback, both positive and constructive, as you start to change?

The final step, 'Do', is a reminder to give your ideas a go. Frontline leaders can be a very effective force in growing and developing team members. By using Energise strategies, particularly during times of change, your team will contribute greatly to the success of the initiative.

Reflection Questions

- How is organisational change usually managed at your organisation? Could things be done differently to increase workforce ownership and energy for the change?
- Are there opportunities for you to support or back-up the initiatives being communicated downwards from management?
- Under what situations would it be 'safe' to increase workers' autonomy and independence, and under which situations could it be risky or even harmful?

Practice Points

- Recognise that many people find change unsettling and anxiety-provoking – develop empathy for these reactions and seek to reassure through involving workers in decision-making.
- Consider how you can make your decisions more participatory and provide workers with an opportunity to become involved in setting priorities, establishing goals, and developing plans to achieve success.
- Channel safety experiences (e.g., a near-miss, a practical safety innovation) that create an emotional response in you, and use these experiences to create an inspiring and motivating style of communication.
- Make sure that safety strategy is translated down into practical and tangible actions that people on your team can actually do.
- Pass on your safety knowledge and skills to your team members, and also facilitate their own insights into individual strengths and weaknesses (coaching).

References

1. Griffin, M. A., Cordery, J., & Soo, C. (2016). Dynamic safety capability: how organizations proactively change core safety systems. *Organizational Psychology Review, 6*(3), 248–272.
2. Zacharatos, A., Barling, J., & Iverson, R. D. (2005). High-performance work systems and occupational safety. *Journal of Applied Psychology, 90*(1), 77.
3. Mayer, R. C., Davis, J. H., & Schoorman, F. D. (1995). An integrative model of organizational trust. *Academy of Management Review, 20*(3), 709–734.
4. Barling, J., & Hutchinson, I. (2000). Commitment vs. control-based safety practices, safety reputation, and perceived safety climate. *Canadian Journal of Administrative Sciences, 17*(1), 76–84.
5. Whitney, D. D., & Trosten-Bloom, A. (2010). *The Power of Appreciative Inquiry: A Practical Guide to Positive Change*. San Francisco, CA: Berrett-Koehler Publishers.

8

Adapt – Making Mistakes

Chapter Summary

Often, we treat mistakes as problems to be avoided, or worse, violations and transgressions that must be punished. This is human nature. Mistakes make us feel vulnerable, exposed, and incompetent. So the natural response when we make one is to fix it as quickly as possible and hope no one notices (or even sweep it under the carpet). Yet, mistakes are a critical part of learning. Without mistakes, we fail to identify gaps in understanding or weaknesses in our organisational systems that can be proactively rectified. To overcome the hesitations of sharing mistakes and engaging in a learning-focussed discussion, we must create a team environment of psychological safety. It is simply ineffective to tell people they must speak up more. Rather, a context in which speaking up becomes the predominant group norm (implicit rule) should be created. Importantly, leaders can create psychological safety through actions like showing their own fallibility. When a team reflects on practice, speaks up, and takes action to improve, they are resilient, which also makes them safer, both in terms of withstanding threats and disruptions, and also using such experiences as opportunities to improve.

From time to time, we make mistakes. But if you can't beat 'em, learn from 'em. Rather than fighting against errors and mistakes (e.g., waging a 'war' on error as per a campaign we came across in one organisation), we should instead be treating mistakes as a chance to learn and improve. After all, errors are inevitable, systems are imperfect, and people have limited capacities and bounded rationality (that is, their decisions are bounded by brain limitations so only an incomplete understanding of the situation can ever be established).[1] Humans simply can't know every piece of information in every situation. This key insight from Herbert Simon in the 1950s changed much thinking about economics and management, eventually disrupting traditional safety thinking as well.[2]

How could a simple idea have such a huge impact? It all comes back to learning and how we learn from mistakes. Think back to any skill you learned such as hitting a tennis ball, running a meeting, or fixing machinery. No matter how much we read about a topic or watch experts performing well, learning often starts off with shaky attempts and with lots of mistakes and errors. Psychologists now realise that these errors are an essential part of learning.[3] We learn as much from doing things wrong, or even more, as doing things right.

If you're involved in safety management you might be thinking 'but mistakes are not acceptable in high-risk work!' or 'how can a good leader allow mistakes to happen?' And that is exactly why Adapt behaviours are challenging. Mistakes can have tragic consequences in safety-critical work, and there are times when learning through error is not possible. In addition, some of the problems with learning from mistakes are that we are so safe that the frequency of major incidents is very low and also that it is very risky, interpersonally and career-wise, to admit that you have stuffed up or failed. Human nature is such that we are predisposed to conceal and minimise our errors, for the most part. We fear rejection and ostracism from the group – an evolutionary throwback to times when being part of a group meant the difference between life and death.

So, here's the dilemma: making mistakes is often a bad idea, yet mistakes are part of learning. This dilemma is why most people agree that learning organisations are important for safety, yet finding ways to support learning to be difficult. We interviewed (with Dr Laura Fruhen) a number of safety professionals about the key behaviours of safety leaders. Adapt behaviours such as 'learning from mistakes' and 'seeing new opportunities for learning' were the least frequently mentioned behaviours. We also find that Adapt behaviours are the least frequently mentioned when we survey employees of organisations.

Learning from mistakes and errors, then, is difficult for a number of reasons including (a) the situation might be too serious to risk any mistake, (b) there is too much fear associated with mistakes to use them as learning opportunities, and (c) leaders are not sure how to turn a mistake into a learning experience. When the situation is too serious, then we really need to focus on Defend behaviours (see the next chapter). But when the challenge is because of fear or lack of skill, then there is new opportunity for creating better safety. Sure, in many cases the fear is justifiable. One mining organisation we know of has a 'window seat' policy. In other words, if you breach a golden rule you're on a one-way ticket back home. However, this company is creating a culture of fear, driving reporting underground and reducing organisational learning. To better understand Adapt behaviours, we need to look more closely at what makes for a good learning organisation.

A learning organisation is one where both good and particularly bad news is welcomed. Many theories of safety culture began with the core idea that poor safety results from inaccurate or missing information. In these situations, one or likely more 'latent conditions' (precursors to an accident) incubate within an organisation, ready to spring a trap for workers operating at the front line.[4] One of the ways these conditions can be surfaced and mitigated is through paying attention to weak signals from the sharp end of operations. Promoting open communication, a sense of unease, and learning from failure are ways to improve the quality of information flow and therefore safety.

In the military, the importance of reflecting on past performance is well known. The After Action Review is one practice that armed forces pioneered, which is now being adopted by many organisations in high-risk industries.[5] In these reviews, teams identify what their original goals were, how they deviated (either positively or negatively), and most importantly, why. A collaborative and psychologically safe (a willingness to speak up and present an authentic version of oneself to others) team environment ensures that the learning is maximised.

Reflection-on-practice is also a hallmark of an effective practitioner. Asking ourselves critical yet curious questions on our past performance with a view to identifying sources of knowledge and ability that helped us to be successful (or alternatively, gaps that increased our difficulty) is key to reflection. Donald Schon identified two types of learning: single loop and double loop.[6] A single loop learning occurs when we might adjust a practice or try a new technique, but the change is largely superficial. A double loop learning is where we transform and develop new insights, beliefs, and assumptions – our whole way of thinking about a practice changes due to deep reflection.

Think about yourself as a leader, either formal or informal, at your workplace. Answer the following questions:

1. What's a time when you didn't quite achieve an outcome that you were hoping to?
2. What were your original objectives or desired outcomes?
3. What did you actually achieve?
4. Why do you think there was a difference between your original objectives and the actual outcome? What things might you have changed and what things were beyond your control?
5. What will you do differently next time?
6. How will you think differently next time?

As we have already mentioned a few times, psychological safety is the feeling within a team that it is OK to speak up, express ideas or concerns, and generally take an interpersonal risk. Psychological safety is a hot topic, with many studies showing that it is related to productivity, innovation, creativity, team communication, and even workplace health and safety performance. It is an important ingredient to make a team successful and high performing. Psychological safety is critical because it taps into the capacity of individuals to make a difference (connect with Energise chapter). Psychologists often describe the untapped potential people have but don't use in their work and everyday life. This potential becomes available when people speak up and are able to act on good ideas. So psychological safety turns on the tap of ideas, innovations, and new learning.

Yet psychological safety is hard to build, and can be destroyed in an instant. Consider a situation where a leader decides to blame a worker for a near-miss, or ridicules someone who spoke up in a meeting for the first time. Not only is that individual less likely to contribute again in the future, but the whole team receives a clear signal that it isn't safe to speak up. Important safety information may therefore be hidden from view.

Rather than telling people to simply 'speak up', it is much more effective to concentrate on building a team environment where this behaviour emerges naturally. From this perspective, we focus less on the actual behaviour and more on the conditions that shape and encourage behaviour. What this means is that leadership practices like recognising people who do contribute, inviting input through asking effective questions, demonstrating fallibility as a leader (i.e., admitting when you make mistakes), asking for feedback on how conversations and meetings went (with a view to improving them), and building personal connections with staff are all ways to foster psychological safety. Taking the time to talk through past performances with a focus on learning, rather than blaming and punishing for mistakes or problems, even implicitly (e.g., acting annoyed or frustrated with people who have made a mistake), also builds psychological safety.

Consider your current team meetings. Do people go quiet when someone asks a question? Are responses usually limited? Is the conversation typically one way? If you answered yes, then your team may have low psychological safety.

The concept of resilience, as used in the LEAD model, is drawn from high reliability organisations. These are organisations where there are high-tempo operations and things can go dramatically wrong in an instant, but mostly do not. In the 1980s and 1990s, a team of researchers headed out to nuclear power plants and military operations like aircraft carriers to try and determine what was going on in these organisations. They discovered five key ingredients that led to a state of 'mindful organising' – a way of design-ing and redesigning themselves to create superior safety and operational performance.

One of the characteristics that high reliability organisations have that makes them effective is a commitment to resilience. Resilience here means that the organisation or the team can withstand pressures and stressors and can even learn to adapt and change itself so it is able to cope better the next time something goes wrong, or a disturbance arises. Having a commitment to resilience means that the team anticipates ahead of time what could go wrong, using 'what if' thinking, and tries to simulate through talking what the situation might unfold like. They also develop responses to these situa-tions proactively, and practice them, so everyone is well-prepared.

Other researchers and thinkers, like Erik Hollnagel, have lent their thoughts to the idea of resilience. Erik proposes that a resilient person (or team, or organisation) has four capabilities: anticipate (what could happen), monitor (keeping a close watch on operations and where things have gone right or wrong), learn (reflect on the past and incorporate new practices and abilities into your repertoire), and respond (react swiftly and decisively to contain problems before they become major incidents).

Think about the work that your team does. What is one aspect that worries you or 'keeps you awake at night'. Consider any high-risk tasks that the team does where controls may be only partially effective or you know the team engages in some risky practices.

What specifically could go wrong with this task? What hazard or hazards exist? How does the team usually control them? What other controls could be applied?

How should the team respond if a hazardous energy is triggered? What process should they undertake? How could you prepare your team for this situation?

Reflection Questions

- When have you reacted 'badly' to a mistake made by yourself or a member of your team?
- Roughly what percentage of errors do you think go unreported by your team? How might this create safety issues?
- What are some of the team processes currently in place to identify, fix, and learn from errors? Are these processes working?
- When could you introduce team discussion on past performance and encourage them to reflect on what could be done differently in the future?
- Do you think your team is currently resilient? What else could they or you be doing to enhance their resilience?

Practice Points

- Work hard to create and maintain psychological safety – it takes time to build and can be destroyed in an instant.
- The most powerful building block of psychological safety is admitting when you don't know or have made a mistake yourself, and inviting input from people in your team without repercussion.

- Use regular opportunities to reflect on past performance with your team, using structured discussions like After Action Reviews.
- Remember to ask workers for feedback on how reflection discussions and learning processes feel to them, and be open to making changes so these activities more closely meet people's needs.
- Practise 'double loop learning' after incidents – what assumptions or beliefs seem incorrect or misleading after the incident, and what should change about the way people think about work and safety moving forward?

References

1. Dekker, S. (2017). *The Field Guide to Understanding 'Human Error'*. Boca Raton, FL: CRC Press.
2. Simon, H. A. (1991). Bounded rationality and organizational learning. *Organization Science, 2*(1), 125–134.
3. Keith, N., & Frese, M. (2008). Effectiveness of error management training: a meta-analysis. *Journal of Applied Psychology, 93*(1), 59.
4. Reason, J. (2016). *Managing the Risks of Organizational Accidents*. Abingdon, UK: Routledge.
5. Morrison, J. E., & Meliza, L. L. (1999). *Foundations of the After Action Review Process* (No. IDA/HQ-D2332). Institute for Defence Analysis. Retrieved from https://apps.dtic.mil/docs/citations/ADA368651.
6. Argyris, C. (1977). Double loop learning in organizations. *Harvard Business Review, 55*(5), 115–125.

9

Defend – High Risk

Chapter Summary

Traditional safety techniques and methods based on reducing or eliminating negative events to 'reasonable' or 'acceptable' levels have allowed dramatic improvements in safety performance. Targeting negative events like injuries and major accidents allowed workplaces to identify what went wrong, and why, and learn from these experiences to improve safety. Typically, preventing negative events draws on the ideas of Taylor, as previously discussed. Standardised procedures are used as a way to prespecify the one best way of doing things. And this is effective when hazards are well known, work is predictable, and conditions are stable. A big problem in organisational safety is the blanket application of Taylor's ideas, rather than a focussed approach that concentrates only on high-risk situations. The LEAD model recognises this distinction and suggests that Defend, which creates a context of stability and prevention through compliance, is best used in high-risk settings. To show Defend, a leader draws attention to risk (creating a sense of unease and vigilance), appropriately monitors work and safety standards, and encourages accountability for safety through fair and just methods.

When work is high risk, a leader's focus must switch to creating a prevention mindset with a strong stability emphasis. This means that workers are clear about their obligations and duties, performance is monitored, and responsibilities are clearly defined and upheld. High-risk work is best managed through traditional methods of safety, termed 'Safety-I', that concentrate on prespecifying the best way to complete the task, slowing down the job where required, and emphasising safety requirements. In effect, safety is clearly prioritised over other goals like efficiency and productivity.

Traditional methods of safety tend to focus on avoiding what goes wrong. Safety is defined by its absence – the reduction of risk to an acceptable level.[1] Defend behaviours have much in common with traditional safety methods because they are concerned with avoiding serious errors and safety risks. Defend behaviours that ensure people are following safe procedures and know about risks are also essential for accountability requirements. Accountability is both a legal and a moral responsibility, and can apply to an individual or a whole organisation. The behaviours of the Defend dimension are a key way of building accountability and showing how accountability is being managed in your organisation. Many stakeholders require clear indicators of accountability for safety including employees, managers, and safety regulators.

Defending is critical when risks are high, and the consequences are severe such as when working at heights or in hazardous environments. But even when vigilance for problems is essential, don't spend 'all' your time with Defend behaviours. Our research shows that Defend is most effective when combined with other LEAD behaviours such as encouraging learning, clarifying goals, and presenting an inspiring vision of the future. People become demoralised when a leader is constantly looking out for problems and never providing supportive feedback.

Defend behaviours such as looking for errors are even more effective when combined with other LEAD behaviours. In one of our studies, we investigated what leadership behaviours encouraged Navy sailors to be most vigilant for mistakes and errors. Unsurprisingly, the least vigilant sailors were those who had leaders who showed very few Defend behaviours. However, more surprisingly, the most vigilant sailors were not those with leaders who constantly showed Defend behaviours such as scanning for risks. What really encouraged vigilance was a combination of Defend and Energise behaviours such as empowering sailors in their team. It turned out that those leaders who focussed only on Defend behaviours were only moderately more effective than leaders who showed almost none of these behaviours. The combination of Defend with other LEAD behaviours was an important insight that we continue to explore.

This example highlights why traditional methods of safety management were not always effective, but also highlights that these methods should not be abandoned.

One of the initial reasons for developing the LEAD model was inconsistency we found between basic safety management and the influential theory of transformational leadership. While traditional safety leadership emphasised vigilance and compliance, promoters of transformational leadership tended to view these behaviours as irrelevant or even harmful aspects of transactional leadership.

How could something essential for safety be viewed in this negative light? The problem lies with a behaviour that was called 'active-management-by-exception' (or AMBE) within the area of transactional leadership. The behaviours of AMBE included pointing out errors and mistakes made by workers. However, it was assumed the leaders who performed AMBE behaviours did so *all* the time. Now, constantly looking for mistakes and correcting others is surely going to be irritating and is not a good way of leading. However, in dangerous situations it can be quite appropriate. The problem was thinking about leadership behaviours as a constant style rather than actions that best fitted a particular situation.

For these reasons, traditional safety practices and modern leadership approaches drifted apart from one another. The LEAD model brings them back together to ensure that good leadership is also good safety leadership.

Traditional safety management and leadership practices have also come under some criticism as modern approaches to leadership attract increasing interest among safety researchers and practitioners. However, the lesson of the LEAD model is not simply to abandon traditional methods. They are still the best methods to achieve safety in high-risk work where hazards are clearly identified and well known. This is because when high-risk jobs are involved, the potential consequences for harm and loss are severe, so a team must be focussed on complying with the best ways that the organisation has found to deal with and manage the risks involved.

Defend behaviours that encourage compliance and vigilance don't mean that the team engages in 'mindless' or surface compliance. A researcher colleague of ours, Dr Xiaowen Hu, developed a model of compliance that contrasts surface level activities like 'ticking the box' on safety with more helpful 'adaptive compliance' where rules and procedures are applied with purpose and depth to meet the local situational demands.[2] Surface compliance can actually be harmful, especially if people are more focussed on showing others they are doing the 'right' thing rather than thinking about the reasons for compliance. Although it might look like people are acting safety, they might miss important risks and fail to respond to unexpected dangers. On the other hand, adaptive compliance means that the team is using procedures as a resource and guide for action, engaging in deep thinking about how to best apply or use them, and actively scanning for hazards in their environment.

But just how do we create adaptive compliance? Well, fortunately, displaying the LEAD behaviours has been shown to predict adaptive compliance. In particular, within the Defend quadrant, a leader will use practices like drawing the team's attention to risk and creating a healthy sense of unease, monitoring work-as-done and ensuring alignment with work-as-imagined, and creating a team culture of fair and just accountability.

A healthy sense of unease is another ingredient shown by a high reliability organisation. Emotions and feelings, although very understudied, play an important role in driving our perception and assessment of risk, as well as our safety behaviours. If we feel complacent and content towards a hazard, our assessment of risk will be much lower than if the hazard induces fear or anxiety. A researcher by the name of Slovic found that there are some predictable characteristics of hazards that affect our assessment of risk, sometimes dramatically.[3] For example, we are more likely to judge a risk as more severe and fear-inducing if it is a novel or new technology, as we are unfamiliar with the hazardous energies involved, if it is man-made rather than naturally occurring, and if it affects children rather than solely adults. So, there is much evidence to suggest that people don't consistently rely on rational and logical methods to evaluate risk.

In short, risk exists both 'out there' in the environment (in the form of danger) and 'in here' within our heads. Risk is both objective (based on facts) and subjective (based on thoughts and feelings). To create a team setting where risk is appreciated, leaders must combine this information and foster a healthy sense of unease. Only then will risks be given the attention warranted by their significance.

According to Dr Laura Fruhen, unease is thought to operate a bit like an inverted-U: too little unease and we become complacent about risk; too much unease and we become paralysed or unable to make effective decisions and assessments towards risk.[4] The sweet spot is somewhere in the middle.

In some organisations, people have literally 'forgotten to be afraid'. Incident-free periods are desired and celebrated. People have become comfortable and complacent about risk. It is not until the next safety incident actually occurs that there is a realisation things weren't quite what they seemed.

In high reliability organisations (HROs), times where things *don't* go wrong are actually anxiety-inducing. There is almost an expectation that things can and will happen, and that periods of quiet mean that something is brewing. In such situations, leaders will actively seek out discrepant information to more accurately gauge the level of risk. Are people just not reporting? Are we missing something important? Are we paying attention to the wrong things? These are all questions that an effective safety leader will ask, rather than putting on the 'zero incidents barbeque' for their staff.

To create a sense of unease and a better appreciation of risk, leaders can do a few things.

First, they can use 'safety imagination'. Think of this like using your capacity for 'what if' thinking. It's using our uniquely human ability to project forward through time but with a safety-specific flavour. What could happen over time? How could this situation deteriorate? What else could go wrong?

Another feature of chronic unease is pessimism and tendency to worry. This characteristic draws on a leader's motivation to consider the worst-case scenario and doubt that everything will turn out OK. This is not to say that an uneasy leader unduly criticises or is harsh and negative overall; rather, they respectfully bring a dose of scepticism to the situation and challenge people to think about what could go wrong. It is sensing that control measures might not be 100% effective, which motivates them to check integrity and monitor effectiveness. In a related batch of research, Dr Stacey Conchie found that safety is highest when there is a combination of trust *and* distrust – that we don't necessarily rely on good faith and an unbridled sense of optimism when dealing with safety.[5] Some trust is helpful for cooperation and openness of communication; yet, for safety, some distrust is also helpful and functional. In other words, cultivating a lingering sense of doubt that everything is under control.

Vigilance is the final aspect of chronic unease. Think of vigilance like 'having your antennae up' around safety-critical activities. It's about directing and focussing attention on the task at hand, minimising multitasking, and concentrating on the present moment. In this way, vigilance is similar to mindfulness. Mindfulness is usually thought about as a form of meditative awareness, something that is only relevant to monks and gurus. Yet, following Prof Ellen Langer's research, mindfulness can also mean 'a flexible state of mind in which we are actively engaged in the present, noticing new things, and sensitive to context'. This means we are focussing on the present moment, eliminating distractions, and being aware of how events are unfolding in light of the situation that unfolds around them (i.e., we ask ourselves why the situation is happening and what this means). So, mindfulness is probably a more helpful concept than vigilance, which usually just means being alert and ready for action.

Again, we remind you that these strategies should not be applied all the time as that can be self-defeating in various ways. Using the other dimensions of LEAD will give meaning and energy to the Defend strategies and will help avoid complacency, exhaustion, and even cynicism about a leader's role. These comments also suggest a more general need to consider variety in the way you act as a leader. Too much emphasis on a single behaviour or a single LEAD dimension eventually 'wears out' the effects of good leadership. Not only is it important to think about adapting your behaviour to the risks of a particular situation (see Chapter 8) but it is also important to think about ensuring a variety of behaviours.

Reflection Questions

- How are people held to account for safety in your organisation? Does this promote learning and healing, or does it emphasise more hurt and punishment?
- How do you typically increase your team's vigilance and sense of uneasiness towards risk? Does your team lean too much towards complacency?
- What percentage of your team do you think engages in superficial or surface compliance? How could you influence them to swing more towards deep and meaningful compliance?
- How can you use team discussion to create a richer and more shared understanding of risks in the work? Think about how you currently run prestart discussions (or similar pre-work activities).

Practice Points

- Treat risk as a combination of both subjective *and* objective elements – the danger is 'real' in the sense that hazards exist in the world, but their risk is best treated as subjective or existing in our heads, so it must be discussed and debated before work commences.
- Consider when adaptive compliance might actually be safer than blind or mindless rule-following.
- For safety, balance trust with distrust – don't take people at face value when high-risk hazards are involved; ensure that you check and probe for confirmation that controls are in place and effective.
- Maintain a constant level of unease across your team so that they don't forget to be afraid – use stories about incidents, safety bulletins, and discussions about near misses and problems.

References

1. Hollnagel, E., Wears, R. L., & Braithwaite, J. (2015). From Safety-I to Safety-II: a white paper. *The Resilient Health Care Net: Published Simultaneously by the University of Southern Denmark, University of Florida, USA, and Macquarie University, Australia.*
2. Hu, X., Yeo, G., & Griffin, M. (2018, July). Safety compliance re-examined: differentiating deep compliance from surface compliance. In *Academy of Management Proceedings* (Vol. 2018, No. 1, p. 15338). Briarcliff Manor, NY: Academy of Management.
3. Slovic, P., Fischhoff, B., & Lichtenstein, S. (1981). Perceived risk: psychological factors and social implications. *Proceedings of the Royal Society of London. A. Mathematical and Physical Sciences, 376*(1764), 17–34.
4. Fruhen, L. S., & Flin, R. (2016). 'Chronic unease' for safety in senior managers: an interview study of its components, behaviours and consequences. *Journal of Risk Research, 19*(5), 645–663.
5. Conchie, S. M., Taylor, P. J., & Charlton, A. (2011). Trust and distrust in safety leadership: mirror reflections? *Safety Science, 49*(8–9), 1208–1214.

10

Safety Culture, Culture Safety, or Culture for Safety?

Chapter Summary

Much has been said about safety culture, but unfortunately, this work has done little to clarify things, especially for practitioners in the field. Safety culture is thrown about but undoubtedly means a whole host of different things. Recently, academics have started to argue against the ongoing use of the term 'safety culture' and instead speak about organisational culture. The safety part then becomes a lens or viewpoint that we apply to the organisational culture, evaluating whether the current culture is either helpful or hindering to the goal of staying safe at work. Understanding the different approaches to safety culture research brings further clarity and a pathway forward: in this chapter we discuss three different streams of work on safety culture. Future ways of thinking about safety culture, namely, as an organisational culture *for* safety (focussing on what culture *does* rather than what it *is*), are also discussed in this chapter.

For decades, debates have covered what safety culture is and isn't and even whether safety culture exists at all. There is some truth to these debates – the term 'safety culture' is a poorly labelled term. Groups have cultures, not things like safety. A group's culture could be reflected in common beliefs about values of individuals; traditions, customs, and rituals around significant events; or indeed any feature that makes a group of people into a 'group'. Once you scratch the surface, safety culture can mean just about anything.

Safety culture has even been described by some researchers, like Sidney Dekker, as the 'new human error', because it becomes a label that blames people further up the chain.[1] Safety culture is easily misused and abused. It can in fact be used to blame groups of people such as 'workers with a poor attitude' or blame undefined aspects of safety management systems for accidents. In the same way that the term 'human error' is used to blame operators, and ultimately isn't all that helpful as an explanation for safety problems, safety culture can also be problematic. It is ill-defined, messy, and means so many different things that it ultimately becomes unusable.

So how do we recover the term and salvage what we can? Well first, when the term 'safety culture' is used, perhaps it is better to think about it as the lens or perspective we talk when analysing and interpreting the broader organisational culture. Organisational culture just exists, and the objective shouldn't be to judge but to understand. This approach is a throwback to when organisational culture was originally developed, and the objective was on studying groups to understand patterns of meaning and behaviour, without necessarily casting judgement on whether it was good or bad.

Over time researchers have realised that it is difficult to study values in different cultures without imposing one's own values on what is being observed. Safety, of course, comes with assumptions about values and beliefs. So, whenever we start putting safety in front of things – safety science, safety management, and safety culture – we are activating and applying a host of different beliefs, values, and judgements. When we say 'safety culture', we have a template or model in mind of what good looks like, and we apply that template to the current organisational culture to determine how far away from an ideal we actually are.

Now this treatment of safety culture causes us to do all manner of strange things. We turn beliefs and perceptions into numbers, assign ratings and numbers, and place organisations on culture 'ladders' – telling them how 'pathological' or 'generative' they are.[2] In reality, this is a form of abuse to the term 'organisational culture'. It reduces a complex social system to a single number or a category. We lose important information through this process, and really water down a concept that is so much richer and nuanced. Sometimes, managers think they have 'measured' the safety culture through this approach and proudly show off their certificates and audit reports (while workers continue to struggle with completing everyday tasks safely, and lament the often-watered-down survey or audit reports they contributed to).

In fact, studies have found that surveys of safety culture and a related concept, safety climate, have inconsistent predictive relationships with safety outcomes.[3] In reality, the relationships are much more complex than simple cause and effect. And at best, safety culture only predicts safety incidents a short distance into the future (about three months). Further, there is a complex relationship between culture and outcomes like incidents. Safety culture explains changes in incidents, and then incidents also explain changes in culture over time.

Given these challenges, how can we summarise the safety culture field? With so many articles, papers, and books, it becomes a difficult job to simplify all this writing into a few categories. Fortunately, researchers have already done so, and people like Frank Guldenmund[4] and Jason Edwards[5] have put forward their ideas for a three-component approach to making sense of safety culture.

The first area of safety culture research focusses on beliefs and meaning. Also called the 'academic' approach, this notion of safety culture is mostly concerned with understanding how people think about and make sense of safety in organisations. It uses methods like ethnography – in-depth interviews and observations – to try to infer or appreciate what is going on inside people's heads. Culture is also treated as something that is divergent or different throughout an organisation. In other words, it doesn't make much sense to talk about one dominant safety culture, but rather, we should try to identify and understand subcultures (teams, departments, professions). This approach provides a rich and detailed understanding of all the different views and ideas about how safety is understood within an organisation. But it is ultimately subjective and driven by the researcher's judgement and interpretation. Take a look at the below statement from a safety culture interview as an example:

[Safety culture has] improved. Probably 12 months ago things were swept under the carpet, it still happens; sometimes people don't want to put things into the system. We have felt the pain of people not reporting. They just think it looks bad. We keep screaming – if we need money or support, the incidents need to be reported. [The phrase] 'Safety costs us money' is still alive and well.

There are multitudes of different ways that this statement can be interpreted. Does the problem lie with the workers' 'failing to report'? Is it a motivational issue? Is it instead a system problem, whereby the reporting software is ineffective and problematic? Should we take a more holistic perspective and investigate how management communicates and deals with reported incidents? As quickly becomes apparent, one criticism of this approach is that it can lead to more studying of culture, rather than any practical solutions or ways forward to improve.

Where does this view of safety culture originate? Well, one key figure is Barry Turner, a sociologist, who in the 1970s put forward the idea that beliefs and shared patterns of meaning and understanding in an organisation can hide signs of disaster.[6] He came to this view after analysing many different organisational accident reports, and noticed that, in most cases, the precipitating events had been lurking or 'incubating' for months, even years, prior to the event. In such organisations, decision-makers had become blinkered to the threats, instead investing their safety efforts on the so-called decoy phenomenon. For example, concentrating on the reduction of person-related safety like slips and trips, at the expense of ultimately more serious hazards like process safety deficiencies. For a range of reasons, the signs of impending doom become masked by the organisational culture. If you are familiar with James Reason's Swiss Cheese model,[7] a similar idea is the notion of 'latent conditions', which are issues that increase risk (e.g., a management decision or inaction) at the blunt end of the organisation (away from the operational work), that lurk and contribute to a disaster in the future.

One of the criticisms of the academic view of safety culture is that it does little more than encourage further study and investigation. It is sometimes unclear how to address the issues except looking in hindsight (i.e., after a disaster has occurred). However, there is likely to be some benefit in conducting detailed descriptions of the safety culture and sharing insights. Doing so may highlight gaps in understanding or in the treatment of risk, resulting in better decision-making. It is akin to holding up a mirror to the organisation and offering them an alternative view derived from the grass roots workers, and indeed, from all levels including leaders. Later in this chapter we offer a possible path forward for the academic view, making it more practical.

Another way of looking at and managing safety culture is by asking people to judge how various safety activities are carried out. This approach almost always involves survey-based methods where people answer many standardised questions. Perhaps the most popular and scientifically developed concept in this area is 'safety climate', which is how people perceive safety in the organisation. In some ways, it makes more sense to talk about safety climate rather than safety culture. Safety climate is much more tightly defined, is validly quantified through survey measures, and has a more consistent predictive relationship with safety behaviours and incidents.

The analogy often used to describe safety climate is the mood or the value and the importance placed on safety in an organisation. People evaluate the safety climate by rating the frequency and effectiveness of multiple policies, procedures, and practices. Safety climate within a team can also be quite different to the overarching safety climate across the organisation. Safety climate differs in this way because team leaders and coworkers have discretion over what and how they implement safety activities. In contrast to a strongly committed management team, a team leader may instead emphasise production over safety, creating a less positive safety climate within their team. This inconsistency creates confusion and uncertainty for team members – what should they prioritise? Is the organisation really committed to safety like they say? Safety performance is likely to suffer as a result of this inconsistency.

Now, safety climate has changed considerably over the decades it has been investigated. Most recent, and also when the LEAD model started its journey, is the idea that different 'flavours' or types of safety climate could be established within an organisation. One distinction made by Jeremy Beus and colleagues is between prevention and promotion climate.[8] If you remember, this is one of the axes that define the LEAD model. Now, most, if not all, existing measures of safety climate do not make this prevention/promotion distinction, and tend to emphasise prevention. This means the questions in safety climate surveys tend to concentrate on the effectiveness of safety procedures (which concentrate on prevention), supervisor's corrective feedback (driving compliance), and the team members' support for safety (which is typically framed as the 'absence of negatives' or preventing harm and loss).

The distinction between prevention and promotion, and as per the LEAD model, also between stability and flexibility, is important when considering the effect of leadership practices on the overall mood or climate towards safety within a team. Leader practices that emphasise growth, nurturance, and achievement needs towards safety are likely to foster a 'promotion-oriented' safety climate, which in turn creates the perception that proactivity and essentially going beyond the minimum is valued and expected. Consequently, workers are more likely to speak up and voice improvement ideas, share concerns, and generally do more than what is required. On the other hand, leadership practices that emphasise the dangers posed by hazards, the importance of compliance, and issuing corrective feedback are likely to create a temporary prevention-oriented safety climate. The leader therefore shapes and moulds the perceptions and expectations of workers, triggering specific types of safety behaviour as a result. Through their practices, it is thought that leaders create a strong situation in which workers receive signals about what is being prioritised and encouraged, with matching behaviours likely to result.

Getting back to the measurement aspect, surveys are the number one tool used to measure safety culture (via the safety climate). From this perspective, safety climate is seen as a momentary snapshot of the underlying safety culture, or a superficial manifestation of the culture at a point in time. It is incomplete because it only taps into perceptions (what people see and hear in the organisation), and not into their deeper attitudes, beliefs, and assumptions. However, the two concepts go together quite nicely: investigations of safety culture should aim to describe and explain, whereas investigations of safety climate should aim to measure and evaluate. The culture just 'is', and shouldn't be described in terms of good or bad. If anything, it can be described as functional or capacity building; the culture produces or does something that is desired/undesired by the organisation. In contrast, safety climate is by its very nature 'normative'. This term means it is the best of what is currently known about a topic, a template or an ideal against which an organisation can be compared and contrasted. Safety climate measurements give us clues about what we should change and how we should change it, whereas safety culture discoveries aim to describe and provide the reason 'why' safety is done the way it is currently in the organisation.

What this means is that measures of safety climate change as our understanding and beliefs about safety also change. Safety climates that focus exclusively on compliance are rapidly being replaced by a broader understanding, whereby workers also need to demonstrate citizenship, proactivity, and adaptive behaviours (not just following the rules). As Safety-II and other ideas become more widely known and applied, we are likely to see a new wave of safety climate research that develops new measures according to these ideas.

The third way of looking at safety culture has been called the 'practical' method. This approach is most closely described by Patrick Hudson's 'ladder' model (you might recognise it more as DuPont's Bradley Curve), which in turn was informed by Ron Westrum's typology of safety cultures.[9] The central idea of the practical safety culture model is that organisations move through or at least can be categorised into one dominant maturity level. With labels such as 'pathological' and 'counterproductive', it is no wonder that such maturity models can leave a bad taste. They are immensely popular across industry, but ultimately lack evidence of their effectiveness.

Maturity models for safety culture are popular for a few reasons. They enable both qualitative and quantitative measurements to be combined into one overall metric (i.e., a maturity level). They also give organisations a road map around which to base their safety journey. They give clues about what's next, and the 'how' of safety culture improvement. And importantly, they are intuitive and are often easy to explain. Finally, they give the company a target to work towards ('we have a calculative culture and we want to move towards a proactive culture'). There are many benefits carried by the practical approach to safety culture.

There are also several risks or downsides to using a maturity model. This approach reduces the complexity of culture to a category and generic description. We know culture is much richer and diverse than a single overall number or level. Treating culture in this way ignores or sanitises the rich data that emerges from data collection. Subcultures and other differences, which hint at diversity within the organisation, are ultimately ignored, leading to a sense of disempowerment and frustration among workers – 'management haven't listened to us'. Relatedly, maturity models dramatically simplify the situation, perhaps at the peril of the organisation. High reliability organisations recognise the limitations of simplifying things. Instead, they refuse to simplify. They embrace complexity and approach it to understand the small signs and weak signals that would otherwise be overlooked.

So, with all these problems and pitfalls, what can be done practically? Well, rather than getting wrapped up in what safety culture *is*, an alternative is to consider what safety culture *does*. Therefore, a more appropriate term would be a culture *of* safety, or even a culture *for* safety. Saying 'a culture for safety' combines multiple perspectives on safety culture together in useful ways. In other words, the academic view provides a rich description of *what* the culture is at a point in time (as well as where it has come from and where it could potentially go), the practical view provides guidance on *where to* next, and the safety climate view provides a way of tracking *change* over time.

Calling it a 'culture *for* safety' highlights how the organisational culture is contributing to (or reducing) safety capability. It is functional and *doing* something, not just existing, either as a level of maturity or a number. That way, we explicitly consider and model how the culture is affecting the organisation and its people. Mapping culture then becomes an exercise in understanding how it is raising or reducing the overall capacity of the organisation to produce safety for its people.

Importantly, under this view, safety culture then becomes the net sum or the product of interactions and relations between different objects. These objects can of course be people, but also inanimate things like equipment, processes, procedures, and systems. Too often, we emphasise the human component of the culture (e.g., the deep-seated beliefs and values) and fail to acknowledge the role of 'artefacts' such as physical tools and equipment (among other things). An object will have just as much a role over the culture for safety as people. When we treat objects with the same importance as people, we open up possibilities for systematic change. It is very difficult to change people's beliefs and values, if not impossible, but it is much easier to change the conditions or context in which they work (e.g., the system of work, structures, etc.), which in turn may produce the outcomes and capability that we desire. We worry less about changing the way people think and more about understanding why they think the way they do and identifying ways of tweaking the organisation to create the performance we desire. We can redesign procedures, change policies, implement new structures, and upgrade/provide/replace equipment, which, over time, will change the culture and create new stories and meaning in the organisation.

Another benefit of this new take on safety culture is the potential to map out both intended and unintended effects. Through gathering stories and other meaning-making activities within organisations, we can appreciate how people make sense of actions, inactions, and the effects of various objects in the workplace. Stories give us a window into the culture by showing us how people interpret their environment. Sometimes, the interpretation may contribute to reduced safety capability (e.g., a particular tool is banned from site, leading to even risker work practices) or produce unintended effects (e.g., a deteriorating physical workplace results in workers losing trust and faith in management, but unexpectedly, in improved safety performance as they become extra cautious and vigilant to threats).

Let's review some of these ideas through a case study. We assisted an organisation to examine their culture for safety through a program of onsite interviews. We met with people at all levels, but predominantly workers drawn from operational positions. A range of questions were asked, including:

- What do you wish that senior management knew about your job?
- What's a past event here that had a positive/negative effect on health and safety? Who was involved, when did it happen, and how did this event affect you or others?
- What's the official line about health and safety at this company? What's the reality of safety like here?
- What does the 'ideal' safety culture look and feel like here?

Many other questions were also asked, but these give a taste. Analysing the answers to these questions across people, we mapped the culture that emerged. Many aspects of the current culture *expanded* safety capacity at this workplace:

- Visible and engaged leaders spending time on the floor.
- A move towards coaching and development around safety.
- Visible safety actions by management to improve the status quo.
- Genuine trust and openness starting to form between leaders and workers.
- Stories of innovation involving contractors and permanent employees working together and persisting through difficulties.
- People taking responsibility for their personal safety.
- Valuing of expertise.
- Safety system seen as credible and in place for the right reasons.

However, we also identified a number of cultural aspects that *reduced* or *contracted* safety capacity:

- The ageing workplace creates a sense of unease and danger – what could fail?
- The backlog of health and safety concerns, and a sense that not enough action is being taken to clear them.
- Keeping quiet about mental health due to stigma.
- Focussing too heavily on statistics, particularly lagging indicators.
- Feeling overwhelmed by a new lock-out-tag-out system that was inadequately supported.
- A history of being punished by safety, creating some hesitation to engage with the safety team.
- Unfair distribution of workload following a restructure and downsizing.

We also identified aspirations for the future – a description of where people from the floor wanted the organisation to head towards in the future regarding health and safety. These aspirations include:

1. Feeling safe: People feel safe, trusting that risks are effectively managed and the workplace is maintained to the highest standards.
2. Owning, innovating, and collaborating: People take ownership of safety and are willing to resolve safety issues together.
3. Safety is ingrained: Safety is invisible because it is a fundamental part of what we do.

After discovering this information, we presented workshops to people from the floor, which helped us to make even greater sense of the data. We also encouraged participants to disclose and document key health and safety stories that they think have shaped the organisation's culture. All this extra information provided valuable context and further insights into the culture for safety.

The point of showing this example is to highlight the rich themes within an organisational culture that can both detract from and contribute to safety capability. These themes also go far beyond traditional ideas about a 'safety' culture. HR decisions, change management, job design, and many other non-safety events ultimately cross over into the safety space. Leaders must recognise that their decisions outside the immediate safety area can have dramatic influences over safety performance. So, we argue it is time to do away with the term 'safety culture' and move towards 'culture for safety'. This change allows us to consider how the broader organisational culture contributes to safety. We remove the lenses and filters that safety culture creates, and embrace a much bigger, and ultimately more effective, understanding of culture and its role in creating safety capability.

Reflection Questions

- When you hear the term 'safety culture', what definition comes to mind? How does it fit within the different streams of safety culture research discussed in this chapter?
- How might the Safety Differently (Safety-II, New View etc.) movement change our definition and understanding of safety culture?
- Why might it be more effective to examine the organisational culture as a whole rather than concentrate on a much narrower part of the culture, that being safety culture?

Practice Points

- Organisational culture is like trail mix: To gain an accurate appreciation of what it is and how it contributes or detracts from safety capability, we need to appreciate its diversity and resist the temptation to oversimplify.
- Culture is like a Petri dish: It is made up of a multitude of different things, animate and inanimate, each with different intensities and levels of influence over the overall culture.

- Culture is like a network: It's a distributed pattern of relationships and effects spread across an organisation; culture exists regardless of the hierarchy.
- Culture is an open system: It will be affected by and will also affect things outside the organisation (consider the role of regulators, market forces, competitors, contractors, etc.).
- Culture is formed and reformed through stories: When events happen in organisations, people form interpretations and often share these with others. When interpretations are shared as stories, they become 'legends' and shape the culture. Stories become a way of teaching newcomers about how to behave and think in the organisation, and of passing on important messages and meaning to colleagues. When a story is told and is accepted as truth or fact, and remains unchallenged, it spreads and creates the culture dynamically.
- Culture is about capability: It constrains some behaviours and enables other behaviours, and ultimately, produces a level of capability to act safely or unsafely.

References

1. Dekker, S. (2019). *Foundations of Safety Science: A Century of Understanding Accidents and Disasters*. Abingdon, UK: Routledge.
2. Hudson, P. (2007). Implementing a safety culture in a major multi-national. *Safety Science, 45*(6), 697–722.
3. Antonsen, S. (2017). *Safety Culture: Theory, Method and Improvement*. Boca Raton, FL: CRC Press.
4. Guldenmund, F. W. (2010). (Mis) understanding safety culture and its relationship to safety management. *Risk Analysis: An International Journal, 30*(10), 1466–1480.
5. Edwards, J. R., Davey, J., & Armstrong, K. (2013). Returning to the roots of culture: a review and re-conceptualisation of safety culture. *Safety Science, 55,* 70–80.
6. Pidgeon, N., & O'Leary, M. (2000). Man-made disasters: why technology and organizations (sometimes) fail. *Safety Science, 34*(1–3), 15–30.
7. Reason, J., Hollnagel, E., & Paries, J. (2006). Revisiting the Swiss cheese model of accidents. *Journal of Clinical Engineering, 27*(4), 110–115.
8. Beus, J. M., Lucianetti, L., & Arthur Jr, W. (2019). Clash of the climates: examining the paradoxical effects of climates for promotion and prevention. *Personnel Psychology.* https://doi.org/10.1111/peps.12338.
9. Westrum, R. (2004). A typology of organisational cultures. *BMJ Quality & Safety, 13*(suppl 2), ii22–ii27.

11

National Culture and Safety Leadership

Chapter Summary

As companies become more globalised, country borders open up, and labour mobility increases, leaders are going to encounter increasingly diverse work teams. National culture has long been studied and has been shown to influence the way people lead, interact with each other, and perform work activities (safely or not). National culture is like a macro layer that sits about the organisational and professional cultures, exerting an effect on thinking and behaviour. By knowing more about national culture, both in terms of what it is and in terms of how this knowledge influences the way leadership should be demonstrated, supervisors can become more effective when leading diverse teams. Drawing on the work of Hofstede, and later ideas from Jason Edwards, in this chapter we discuss how the LEAD strategies and behaviours might be implemented differently to accommodate for national culture. So, rather than going against the grain when it comes to safety and national culture, leaders should instead seek to adapt their own ways of inspiring and motivating others to achieve optimal outcomes.

Leaders in today's modern workplaces must often manage diverse groups, both in terms of skillsets and abilities, as well as national culture. It is helpful to think about national culture like a bubble surrounding an individual. Within this bubble there may be other bubbles, such as professional culture, organisational culture, and team culture. Each of these 'bubbles' exerts a subtle influence on thinking and behaviour. These bubbles are different layers of culture.

Being a part of a society means that people adopt values and beliefs, which in turn shape their behaviour. When a culture exists, it means that multiple people share similar ways of thinking – they have a 'shared mental model' or a template about how the world around them works and how they should interact with it to achieve goals. Most of the time, in fact nearly all of the time, we are blissfully unaware of these different cultures; nevertheless they affect us in deep and complex ways.

There have been many attempts to measure and make sense of national culture. Geert Hofstede[1] is one of the most well-known researchers in this area, however, his work has been extended by a multinational team through the GLOBE project.[2] This project is a multinational survey of cultural practices, ideal leadership attributes, and antecedents of trust across more than 120 countries around the world.

Importantly, experiments involving the theory that the LEAD model operates from (self-regulation) show that the principles apply universally across cultures, however, they are expressed differently. For example, in Eastern cultures, the importance of 'face' (the respect and deference a person can claim from others due to their social status or position in a hierarchy) means that they are predisposed to respond in ways that emphasise 'avoidance' or to minimise any potential harm or loss. Applied to the LEAD model, it may be therefore 'easier' for a leader to achieve compliance with procedures through applying a Defend strategy. In Western cultures, the importance of 'self-esteem' (positive self-perceptions) means that a worker is more easily induced to adopt an 'approach' style of motivation whereby they seek to maximise successes and achievements. This might translate into an easier implementation of the Leverage strategy when applied by a leader.

However, there are a few dimensions of national culture that have special significance for safety[3] and the LEAD model. These dimensions mean that the effectiveness of the LEAD model may differ depending on national culture, provided no changes or adaptations are made to how the practices are demonstrated. The national culture dimensions described by the GLOBE project that are most relevant to LEAD include:

- Uncertainty avoidance
- Power distance
- Collectivism
- Future orientation

Uncertainty avoidance is a culture's effect on someone's tolerance for ambiguity and the unknown, for instance, their preferences or rules, structure, and direction as opposed to autonomy and approaching uncertainty. Leaders dealing with high uncertainty avoidance would likely have an easier time implementing procedures, rules, and structured processes, and achieving compliance from workers. On the other hand, new safety innovations or opportunities to explore novel solutions over and above tried-and-true methods are likely to be more difficult. A leader may have to work extra hard to provide a reassuring environment in which cultures high on uncertainty avoidance feel safe and comfortable to engage in change.

Power distance is the degree of separation or boundary between a leader and their followers. In a high power distance country, workers will tend to defer to supervisors' decisions and feel uncomfortable challenging a leader's intentions or engaging in participative decision-making. In other words, high power distance cultures expect leaders to instruct and use their authority. Many countries, including the UK and Australia, have extensive consultation provisions in their work, health and safety legislation. Dealing with multinational workforces, particularly those with high power distance, may require new methods to achieve the same outcome. For instance, a leader may need to rely on a trusted 'second in charge' or other peer resource that is at a similar status level to the team to carry out workforce consultation. There may also be benefit in asking hypothetical questions rather than being more direct, for example, 'if you were in charge of safety for this job, how would you approach it?'. Critical is to avoid the sense that workers are directly criticising or undermining the leader's approach.

Collectivism refers to the strength of group rules and norms and to identification with the organisation rather than emphasising individuality and uniqueness. In cultures low on collectivism, workers will likely want their 'time to shine' as individuals and will be more motivated by personal recognition, rewards, and performance feedback. In collectivist cultures, workers will tend to be more accepting of group feedback and recognition. They will want to feel that they are contributing to group or collective goals rather than how they are performing as individuals. In collectivist cultures, organisational campaigns and initiatives to improve safety that emphasise how individual actions will help contribute to group success may be more effective.

Future orientation is the tendency to concentrate on past events and the present moment (e.g., immediate production demands) rather than long-term considerations like safety or sustainability. In cultures low on future orientation, people are more likely to be motivated by immediate gains or rewards like production targets/outputs. Cultures high on future orientation will be more likely to deeply consider how current actions contribute to future threats or opportunities, so, for example, they may be more engaged during prestart meetings or other preparatory/planning sessions.

Reflection Questions

- How diverse is your team currently? Do you think it will become more diverse as time goes on?
- Do you find that you naturally adjust or adapt your leadership behaviours to more effectively interact with workers from diverse backgrounds? What exactly do you do that is more/less effective?

Practice Points

- Be mindful of how your leadership actions may come across to different cultures, and take the time to learn how people from different cultures prefer to be involved, consulted, and managed when it comes to safety.
- Consider how your own national culture might bias or influence your ideas about what it means to be an effective leader.
- Be mindful of giving workers the right amount of clarity, certainty, and independence, taking into account personal preferences (which will incorporate national culture).

References

1. Hofstede, G. (2011). Dimensionalizing cultures: the Hofstede model in context. *Online Readings in Psychology and Culture, 2*(1), 8.
2. Chhokar, J. S., Brodbeck, F. C., & House, R. J. (Eds.). (2013). *Culture and Leadership Across the World: The GLOBE Book of In-Depth Studies of 25 Societies.* Abingdon, UK: Routledge.
3. Yorio, P. L., Edwards, J., & Hoeneveld, D. (2019). Safety culture across cultures. *Safety Science, 120,* 402–410.

12

Leading Safety in the Future

Chapter Summary

The changing world of work means that what happened in the past doesn't always translate unchanged into what we should do in the future. The digital revolution has changed how work is done, how people are connected with one another and with computers/machines, and the nature of the employment relationship itself (e.g., the 'gig economy'). These trends influence a workplace like a double-edged sword, creating unparalleled opportunities as well as unprecedented threats at the same time. In this chapter, we focus on three mega trends, automation, data and communication, and the gig economy, and discuss their impact on safety leadership.

Automation is one of the biggest changes that have been made to organisations, particularly in manufacturing and transport, often with dramatic impacts on safety. In the case of mining, automation of excavation, loading, and transport have a positive impact on safety because of increased control over the work process and removing workers from risky environments. Also, automation of complete or parts of a work process can free up human operators to focus on more complex and interesting work, eliminating factors like boredom and distraction (which are often identified as precursors to accidents at work).

As a result, you'd be well placed to think about automation as something that increases safety and might even make safety leadership less relevant and important in the future world of work. But hold up a second. Safety could just be an illusion in this case. Operators who think that automation will 'save them' from disaster might have false assurance – they think they are safe, but in reality, they aren't. Automation also makes a work system much more complex, with often unanticipated interactions. When things fail, operators might also be less equipped to problem solve and maintain safety (the technology is too difficult to understand).[1]

Even though automation is on the rise, people will always be needed because they give a uniquely human capability: adaptation and flexibility.

People, particularly experts, have the uncanny ability to bring a work system back from the brink of failure. And where workers are needed, there will also be a requirement for someone to lead them.[2]

Automation may even introduce health-related demands on leaders of teams. People often use the term 'innovation fatigue', which describes the consequences of too much change in work processes and systems. People become tired and disengaged by constant upgrades and replacements, given they are likely needing to unlearn, learn, and relearn various job skills. Leaders are likely to be put into the position of driving this change while also maintaining work goals like productivity and safety. Leaders will need to inspire and motivate workers through this fog of innovation fatigue, while also encouraging their teams to consider how safety might be impacted. Consultation within teams will probably become even more critical than it is now, with leaders forming a critical role in coordinating and encouraging such feedback to management and other stakeholders.

Not to mention the increased job demands that future work will create on employees. As jobs become more automated, operators will need new and different skills and knowledge. Leaders will need to step up and extend the training given by their organisation, and support workers through the changes that are likely to happen through new technologies and methods of doing work. Leaders may also need to be more aware of worker well-being and mental health, as the workplace becomes more stressful.[3] There are many challenges that future leaders of health and safety will need to overcome.

We are inventing new and innovative ways of keeping tabs on how work and workers are progressing. We monitor people and process with sensors, algorithms, and a host of other technologies. Artificial intelligence is used to help improve work, crunching through this information at lightning speed. Technology also allows more flexibility over how people work, which means virtual team environments are going to increase. Sensors and algorithms can also give workers more real-time safety information. These technologies could help teams to build a shared awareness of the situation more rapidly and detect threats to safety earlier.

At the same time, teams also run the risk of being overwhelmed by data, which takes time to process and make sense of. Have you heard of the term 'big data'? It's a buzzword that gets used a lot but ultimately is not really living up to its promise of improved information for organisations. What happens is that garbage in equals garbage out. In the rush to use technology and data gathering systems, organisations may have taken the plunge too early and created a situation where their incoming data is difficult to make sense of and action. Workers are also likely to suffer from information overload and, therefore, less effective decision-making. Peer pressure to respond to work-related messages could also increase, creating issues like burnout and sick leave. The boundary between work and home could also become even more blurred. For example, research has found that workers who use their smartphone (for work purposes) before bed are likely to have poor sleep, which makes them fatigued and distracted the next day. Virtual teams could also become more socially isolated, as organisations move to a model where people are encouraged to work from home. All of these issues could potentially impact safety, meaning that leaders of the future will need to be across them and ready to respond where needed.

Leaders in the future will need to be savvy with technology if they are to capitalise on the benefits for health and safety. They will need to use technology to their advantage. Leaders will likely have increased access to safety data, including information on their team's safety performance. On the one hand, this should make it easier to be aware of when feedback and recognition should be given, but runs the risk of leaders spending even less time out in the field. Telecommuting and remote work arrangements mean that leaders will need to find new ways of connecting to their teams, forming meaningful relationships, and encouraging the same between other team members separated by distance and computer screens. Again, leaders will need to be vigilant to signs that a team member is struggling or burning out and will need to make sure these psychosocial hazards are appropriately controlled. In general, the balance is likely to switch from physical hazards to the psychological.

Recently, we have seen the rapid rise of the 'gig economy', where people take up temporary job assignments through digital platforms like Uber, Airbnb, and Taskrabbit. The gig economy challenges how we see work and organisations, because these future workplaces will probably consist of only loosely connected individuals.

The gig economy will likely carry many opportunities, and risks, for safety. For example, organisations will probably be able to hire extra resources to monitor and maintain. On the other hand, the gig economy model means that our traditional understanding of a 'safety culture' is going to be challenged. How could such an organisation of independent people create a culture for safety? Also, the gig economy connects workers and customers much more closely than ever, with all the pressures and demands that such a relationship brings. Customers will demand faster and higher quality, tempting workers to perform workarounds and take other risks to get the job done. Workers may be hired on increasingly short-term and temporary assignments, with huge job insecurity and inconsistency in work standards as a result.

From a leadership perspective, we will need to work extra hard in such an environment to offset and even counter the temptation to focus more on driving job performance (e.g., how quickly a task is performed) and less on the intangible leadership activities that foster a positive workplace culture (e.g., teamwork, feedback, positive communication). Organisational culture may weaken as a result, which means that workers will have fewer signals to remind them of how to behave in safety-critical settings. Essentially, organisations will lose the 'soft control' that culture facilitates.

Leaders may also need to become more savvy and more intimately involved in recruitment and selection for safety. If organisations become less connected and more about individuals, there will probably be a greater reliance on workers' skills, knowledge, and personality characteristics to keep them safe. Leaders will need to be across the relevant capabilities that workers need in specific roles to ensure safety is achieved.

Training could also change under a gig economy. At the moment, it is expected, and even legislated in most countries, that organisations will provide workers with formal safety training. In a gig economy, such training could decline or disappear altogether. In addition to online learning opportunities, leaders might need to step in and fill this gap by giving gig workers training (e.g., webinars or other virtual learning sessions). This could mean that leaders themselves receive even less development than they do now. Studies in Australia have shown that supervisors are particularly underdeveloped in the workplace, with few leaders receiving any formal training, or developing in preparation for taking on the role. Leaders will probably need to be more independent and self-directed in seeking out new opportunities to learn, about their general leadership role as well as how to lead and manage safety.

Overall, work is rapidly changing. Technology changes and the nature of work itself – shifting from traditional organisations to more dispersed and temporary employment arrangements – are likely to pose challenges for safety, and for leading teams. The nature of a team will undoubtedly change as well, with more virtual groups connected only via electronic communication becoming the norm. The LEAD model is versatile – it will help you to adapt and adjust to this changing nature of work.

Reflection Questions

- What technological changes are most likely to impact your workplace?
- How can you use leadership to prepare, engage, and inspire your team to create resilience in the face of technological advancements like automation?
- How can you use new technologies to free yourself of non-value-adding activities that prevent you from spending time out in the field interacting with your team?

Practice Points

- Always consult with your team before new technologies or other advancements in work process are implemented – just because it is computer controlled doesn't make it safer.
- Look for ways to keep your workers engaged and attentive to the task when work is automated.
- Consider how your team would react and respond when automated technologies fail – are there ways to prepare for these situations by giving your team additional training or preparation for what to do when things go wrong?
- Keep data in context by discussing with your team and peers, rather than relying on the numbers alone.

References

1. Hollnagel, E. (2018). *Safety-I and Safety-II: The Past and Future of Safety Management*. Boca Raton, FL: CRC Press.
2. Dekker, S. (2019). *Foundations of Safety Science: A Century of Understanding Accidents and Disasters*. Abingdon, UK: Routledge.
3. Ellis, A. M., Casey, T. W., & Krauss, A. D. (2017). Setting the foundation for well-being: evaluation of a supervisor-focused mental health training. *Occupational Health Science*, 1(1–2), 67–88.

Appendix 1: The LEAD Backstory

LEAD began its life as a safety leadership training program developed for a large iron ore miner in Western Australia. At the time, Professor Mark Anthony Griffin and his team were engaged to build and evaluate LEAD training for the organisation.

Fast forward a few years. Thereafter, Dr Tristan William Casey picked up where Professor Griffin and colleagues left off, and along with Professor Andrew Neal from the University of Queensland, began redevelopment and expansion of the LEAD model. A group of 25 safety leadership subject matter experts were interviewed about different safety leadership practices to effectively expand the LEAD model in more detail. Their responses were analysed and then used to develop a pool of survey items that could be used to measure each LEAD strategy. While at the Office of Industrial Relations, Dr Casey engaged with over 50 different companies to trial and test different versions of the survey tool. A final validated version was then developed, and is now offered by the Queensland Government on its Safety Leadership at Work website.

Around this time, Teys Australia was working with Dr Casey to implement the first revised version of the LEAD training package (see Appendix 2). A first modularised version of the LEAD program was then developed. This program consisted of five sessions, each two hours in length. Participants focussed on practical skill development and practised through workplace projects conducted in-between each session.

Thereafter, Dr Casey engaged with a local university and council to develop and evaluate a shorter version of the LEAD program. Just two hours in length, two versions of the program were created – one for managers and one for workers. In total, more than 300 people participated in the revised version of the LEAD training and provided feedback that helped to iterate new versions of the material every time it was conducted.

Other people and organisations, including the Electrical Safety Commissioner and Energy Skills Queensland, have since taken up the LEAD model and training. As the LEAD model continues the gain momentum, there will be plenty more case studies showcasing the great work being done through government, industry, and academic collaborations.

Appendix 2: LEAD Case Study: A Training Implementation at Teys Australia

Teys Australia is a large livestock farming and meat processing company headquartered in Brisbane, Australia. They have over 3,000 employees and a number of processing facilities nationwide.

In 2018, Teys expressed some interest in working with us to develop, implement, and evaluate a bespoke training program based around the LEAD model. The training targeted frontline supervisors, because Teys recognised the important role of this level in encouraging safe working practices 'at the coal face'. Further, in line with other major studies in Australia, Teys acknowledged that many of their supervisors and team leaders had been promoted into these roles with only basic training and support to build advanced communication and leadership skills.

To begin the training development and testing, a small group of 10 supervisors from across a meat processing plant were asked to participate. These people were chosen because of their informal leadership status and influence over others at the site. A draft version of the training was delivered to this group over five weeks – each module was approximately two hours in duration and covered each core LEAD strategy (as well as a general introductory module about safety leadership). Following the draft training implementation, each participant was interviewed and their feedback data were analysed to identify where we could make improvements.

After this step, Teys identified a group of champions across their operations who could act as trainers and supporters of the LEAD program ongoing. The champions were passionate supervisors, safety staff, and professional trainers and human resources advisors drawn from different locations and business units. We trained these champions in methods of LEAD implementation and supported them to start rolling out the training across Teys. In addition to the program content, these champions were given training in learning theory and facilitation techniques over two days.

Results were significant. Part way through roll-out we surveyed leaders and workers to capture their impressions of the training impact. Workers reported that the safety climate (perceptions of safety value and importance) were above the industry benchmark and that they had observed noticeable changes in their leaders' safety practices. Leaders themselves reported that they had found use and benefit in applying the LEAD practices in their workplace.

At the time of writing, Teys was embedding the training by developing a LEAD refresher course (online), incorporating LEAD ideas into standard employee inductions, and implementing key performance indicators relating to LEAD practices and safety behaviours among staff.

Culminating the LEAD implementation at Teys was their achievement of two awards in 2019 from the National Safety Council of Australia (NSCA). In addition to being awarded the 'best safety leadership training', Teys was also recognised with the coveted 'pinnacle' award from the NSCA.

Appendix 3: LEAD Case Study: Robust Risk Conversations at a Water Services Company

The company involved is a major contractor in Queensland who services and constructs water infrastructure. This organisation has approximately 300 employees and performs a variety of high-risk work tasks that ranges from maintenance and refits of sewerage treatment plants through to finding and fixing leaks in water pipe networks across the city of Brisbane and beyond.

The company is a progressive organisation, having launched their 'performance evolution' based on principles from Safety-II and the field of Human and Organisational Performance. The performance evolution treats safety as an integrated part of successful work, rather than an additional layer of arduous requirements and compliance activities. Consequently, they have enjoyed a dramatic turnaround in staff engagement and performance through this humanistic strategy.

In 2019, the company approached us to help them develop a training package and tool that could be used to replace an existing paperwork-heavy risk assessment process with a more dynamic and, ultimately, more effective guided conversation process. Following extensive consultation and co-design with organisational stakeholders at various operational levels (including some safety staff), the LEAD model was identified as a promising framework around which the conversation process could be designed.

The training package concentrated initially on developing advanced conversation and listening skills to support the implementation of the tool. We trained attendees in the art of effective 'humble' questioning, active listening techniques, development of psychological safety, and defusing and de-escalating conflict. At the conclusion of the training we provided a scaffolding that could be used to guide future pre-job (prestart) meeting conversations, covering job setup, innovation, quality improvement, and safety.

Each of these areas was mapped on the LEAD model. Job setup and initiation were covered by questions relating to the Leverage strategy. We encouraged participants to ask and discuss questions such as: What does a 'good' job here look like? What extra equipment or support do we need to be successful? How can we maintain up-to-date and accurate communication throughout the job?

Innovation was covered by the Energise strategy. We provided example questions such as: Is there a smarter/safer/better way to do this job? What ideas do we have to do this job differently? When have we been particularly successful with this job in the past and why?

Quality improvement was covered by the Adapt strategy. The intention of the Adapt-focussed discussion was to stimulate learning and improvement by reflecting on previous occurrences of the same or similar jobs in the past. Questions asked during this discussion section could include: When did we have to improvise or make do in the past? What surprised us or caused problems the last time we did this job? What did we learn when we completed a similar task before?

Safety, in particular, dealing with high-risk hazards, was covered by the Defend strategy. Teams could ask questions such as: What can cause harm, damage, or loss here? How can we manage any risks to acceptable levels? What could cause problems on the job? When and where do we need to slow down the job and take extra care?

Results were promising and showed that the quality and nature of workers' discussions significantly changed after the LEAD conversation process was implemented. Participants were surveyed before and about one month after the training was implemented. Participants had more frequent discussions about all aspects of job success and, overall, felt significantly more inspired to consider innovative ways to do the work and to be extra vigilant to risk in their work environments.

Although this case study involved a heavily adapted version of the LEAD model, it is an example of how versatile and adaptable the LEAD framework can be. Consider how you might be able to innovate the LEAD model in your organisation and use this case study for inspiration.

Appendix 4: Changing an Entire Industry

As part of the Electrical Safety Commissioner's (Queensland, Australia) vision and action plan for 2019–2020 and beyond, the LEAD model was used as a framework to guide the development of an industry-wide safety leadership training package. The electrical industry continues to experience many dangerous electrical events and, unfortunately, some fatal incidents like electrocutions. In his role supporting the Electrical Safety Office (Office of Industrial Relations) and Electrical Safety Board, Mr Greg Skyring is involved in a number of disciplinary hearings following dangerous electrical events that catch the attention of the health and safety regulator in Queensland, Australia. Many of these incidents include a component of supervision deficiency.

To improve the industry as a whole, Mr Skyring, along with Energy Skills Queensland, identified and engaged with just over 20 different electrical contractors of various sizes to participate in safety leadership training based on the LEAD model. Dr Casey supported the design and development of this revised training package, which was delivered to the cohort in mid- to late 2019. In early 2019 the whole cohort was brought together to inform the design of the training. PricewaterhouseCoopers facilitated a full-day design session at Energy Queensland. Industry provided direct input into the creation of the final training package by sharing their ideas and concerns through facilitated collaboration events. Also, the long-term sustainability of the training was discussed and a blended online/face-to-face model was identified as the most suitable approach.

The training package is a brief two-hour introductory session that includes a number of short exercises and reflection points across all aspects of the LEAD model. The aim of the training is to introduce the concept of safety leadership and the LEAD model and to facilitate the setting of a few key safety-related goals. Participants also learn about tools that correspond to each quadrant of the LEAD model, such as After Action Reviews (Adapt) – a structured post-shift or post-project learning process, implementing a just and fair team culture (Defend) and performing safety recognition effectively (Leverage).

The involved companies also participated in a short survey before the training and again after the training, to measure changes. Before the training, the survey informed the development of an action plan. Companies participated in a debrief session that explained the survey results and encouraged business owners and senior managers to come up with targeted improvements.

Feedback from the training sessions was positive, and the follow-up survey was launched in March 2020. The survey analyses found that some changes were apparent in team members' safety behaviours, which could have been due to the improvement in supervisor safety leadership. This information will be used to evaluate and further refine the training. In the future, it is possible that the LEAD training will be incorporated into electricians' professional development requirements, and disciplinary hearing participants will continue to be referred to the program, where appropriate. People interested in participating in this version of the program can contact Energy Skills Queensland.

Index